Springer Theses

Recognizing Outstanding Ph.D. Research

Aims and Scope

The series "Springer Theses" brings together a selection of the very best Ph.D. theses from around the world and across the physical sciences. Nominated and endorsed by two recognized specialists, each published volume has been selected for its scientific excellence and the high impact of its contents for the pertinent field of research. For greater accessibility to non-specialists, the published versions include an extended introduction, as well as a foreword by the student's supervisor explaining the special relevance of the work for the field. As a whole, the series will provide a valuable resource both for newcomers to the research fields described, and for other scientists seeking detailed background information on special questions. Finally, it provides an accredited documentation of the valuable contributions made by today's younger generation of scientists.

Theses are accepted into the series by invited nomination only and must fulfill all of the following criteria

- They must be written in good English.
- The topic should fall within the confines of Chemistry, Physics, Earth Sciences, Engineering and related interdisciplinary fields such as Materials, Nanoscience, Chemical Engineering, Complex Systems and Biophysics.
- The work reported in the thesis must represent a significant scientific advance.
- If the thesis includes previously published material, permission to reproduce this must be gained from the respective copyright holder.
- They must have been examined and passed during the 12 months prior to nomination.
- Each thesis should include a foreword by the supervisor outlining the significance of its content.
- The theses should have a clearly defined structure including an introduction accessible to scientists not expert in that particular field.

More information about this series at http://www.springer.com/series/8790

Andreas Reiserer

A Controlled Phase Gate Between a Single Atom and an Optical Photon

Doctoral Thesis accepted by
the Max-Planck-Institut für Quantenoptik, Germany

Author
Dr. Andreas Reiserer
TU Delft
Delft
The Netherlands

Supervisor
Prof. Gerhard Rempe
Max-Planck-Institut für Quantenoptik
Garching
Germany

ISSN 2190-5053 ISSN 2190-5061 (electronic)
Springer Theses
ISBN 978-3-319-26546-9 ISBN 978-3-319-26548-3 (eBook)
DOI 10.1007/978-3-319-26548-3

Library of Congress Control Number: 2015954985

Springer Cham Heidelberg New York Dordrecht London

Printed on acid-free paper

Springer International Publishing AG Switzerland is part of Springer Science+Business Media
(www.springer.com)

Parts of this thesis have been published in the following journal articles:

- *A quantum gate between a flying optical photon and a single trapped atom*
  A. Reiserer, N. Kalb, G. Rempe, S. Ritter
  Nature **508**, 237 (2014)
- *Nondestructive Detection of an Optical Photon*
  A. Reiserer, S. Ritter, G. Rempe
  Science **342**, 1349 (2013)
- *Generation of single photons from an atom-cavity system*
  M. Mücke, J. Bochmann, C. Hahn, A. Neuzner, C. Nölleke, A. Reiserer, G. Rempe, S. Ritter
  Physical Review A **87**, 063805 (2013)
- *Ground-State Cooling of a Single Atom at the Center of an Optical Cavity*
  A. Reiserer, C. Nölleke, S. Ritter, G. Rempe
  Physical Review Letters **110**, 223003 (2013)
- *Efficient Teleportation Between Remote Single-Atom Quantum Memories*
  C. Nölleke, A. Neuzner, A. Reiserer, C. Hahn, G. Rempe, S. Ritter
  Physical Review Letters **110**, 140403 (2013)
- *An elementary quantum network of single atoms in optical cavities*
  S. Ritter, C. Nölleke, C. Hahn, A. Reiserer, A. Neuzner, M. Uphoff, M. Mücke, E. Figueroa, J. Bochmann, G. Rempe
  Nature **484**, 195 (2012)
- *A Single-Atom Quantum Memory*
  H.P. Specht, C. Nölleke, A. Reiserer, M. Uphoff, E. Figueroa, S. Ritter, G. Rempe
  Nature **473**, 190 (2011)

Parts of this thesis have been published in the following journal articles:

- [illegible]
- [illegible]
- [illegible]
- [illegible]
- [illegible]
- [illegible]
- [illegible]

Supervisor's Foreword

Today's internet is a powerful communication network but faces limitations which can only be overcome with a network based on quantum-physical principles. These will allow us to perform tasks and to interact in ways which are not possible in the framework of classical physics.

The realization of such quantum network is a key challenge for quantum science. It requires stationary quantum nodes that can send and receive as well as store and process quantum information locally. These nodes are then connected by flying information carriers, i.e. optical photons, which facilitate a direct exchange of quantum states between nodes and the generation of entanglement of two or several nodes in the network. An efficient interface between the nodes and the channels is a prerequisite for the scaling of quantum networks to a large number of particles and to long distances.

Against this backdrop, Andreas Reiserer employed an experimental system which was developed in my group at the Max-Planck Institute of Quantum Optics. It involves a single atom trapped in and coupled to an ultrahigh-quality optical microresonator. This approach gives access to single qubit manipulations, long coherence times and high light–matter coupling efficiencies. The first experiments carried out in the course of Andreas' doctoral work have employed the light–matter interaction technique of vacuum-stimulated Raman adiabatic passage. In this way, Andreas and his colleagues generated entangled photons at the push of a button, mapped the quantum state of a photon onto an atom and back, transferred and teleported quantum states between two remote atoms, and entangled two atoms over a physical distance of 21 m, using a 60 m long optical-fibre link. These experiments demonstrated the great potential of atom-cavity systems for the implementation of quantum networks.

The full exploration of this potential, however, required improving the localization and reducing the thermal motion of the trapped atom. To this end, Andreas implemented a three-dimensional optical lattice with high trap frequencies along all spatial directions. Using a pair of Raman lasers, full control over the quantum state of a single atom trapped in a cavity was demonstrated. This enabled the strong-coupling regime of cavity quantum electrodynamics to be reached in a resonator of sufficiently large size to allow unperturbed optical access.

This configuration then facilitated a remarkable series of experiments, all based on an interaction mechanism that was proposed in 2004 by Duan and Kimble. The mechanism employs photon reflection from a resonant cavity and is robust with respect to experimental imperfections. The experimental highlights achieved by Andreas and colleagues include, among others, the non-destructive detection of a single optical photon, a long-standing dream of quantum-optical research since Braginsky's idea from around 1990; the first-ever realization of a quantum gate between a stationary atom and a flying photon; and, not least, the entanglement of an atom with two successive photons.

The experimental techniques reported in Andreas' thesis should in principle also be applicable in other physical systems that can strongly couple different information carriers, such as atom and photon, or spin and magnon. I am therefore convinced that Andreas' achievements will have a broad and long-lasting impact on the evolution of quantum science and technology.

Garching
September 2015

Gerhard Rempe

Acknowledgements

This thesis was prepared in the Quantum Dynamics Division at the Max-Planck-Institute of Quantum Optics. During the last years, I have greatly enjoyed and benefited from working in this inspiring group. In the following, I want to thank all colleagues and friends whose constant support has made this thesis possible.

First, I want to express my deep gratitude to Gerhard Rempe for being my doctoral advisor in the past years, and for sharing the enthusiasm for single atoms and single photons with me. In particular, I want to thank him for his constant and inspiring scientific contributions, and for supporting me and our work in every respect. I also want to thank Stephan Ritter, who supervised the quantum information experiments during my thesis, and whose friendship and scientific input cannot be appreciated high enough. Thanks also for carefully proof-reading this thesis!

During my thesis, I could build on the work of my predecessors at the Qgate experiment: Stephan Nußmann, Markus Hijlkema, Bernhard Weber, Holger Specht and Christian Nölleke. I want to thank them for conceiving and setting up such an exceptional apparatus. I especially enjoyed working with Christian, and I want to thank him for teaching me countless things, such as atom trapping, laser locking and—occasionally—tabletop soccer. In addition, I want to thank "my" Master student Norbert Kalb, who has learnt to operate and advance the experiment much faster than I had thought possible. I really enjoyed our teamwork, and I hope that we can continue with it in Delft!

I also want to thank the whole Quantum Information team for our stimulating discussions, and for facilitating the joint measurements by working day and night. In particular, I want to thank Jörg Bochmann, Martin Mücke, Carolin Hahn, Andreas Neuzner, Manuel Uphoff, Eden Figueroa, André Kochanke, Johannes Lang and Johannes Burkardt.

Many thanks also to the technicians Josef Bayerl, Franz Denk, Helmuth Stehbeck, Tobias Urban and Thomas Wiesmeier for their steady support. And I would like to express my thanks to the whole Quantum Dynamics Group, for the excellent cooperation, the excursions and group retreats, the parties … in short: the great team spirit that we had!

Finally, I want to thank my family and my wife Eva, for their encouragement, constant support and love.

Contents

Abstract

This thesis reports on the experimental implementation of a deterministic interaction mechanism between flying optical photons and a single trapped atom. To this end, single rubidium atoms are trapped in a three-dimensional optical lattice at the centre of an optical cavity in the strong-coupling regime. Full control over the atomic state—its position, its motion, and its electronic state—is achieved with laser beams applied along the resonator and from the side. When faint laser pulses are reflected from the resonator, the combined atom–photon state acquires a state-dependent phase shift. In a first series of experiments, this is employed to nondestructively detect optical photons by measuring the atomic state after the reflection process. In a second series of experiments, quantum bits are encoded in the polarization of the laser pulse and in the Zeeman state of the atom. The state-dependent phase shift then mediates a deterministic universal quantum gate between the atom and one or two successively reflected photons, which is used to generate entangled atom–photon, atom–photon–photon, and photon–photon states out of separable input states.

Chapter 1
Introduction

1.1 Atom-Photon Interaction

The interaction of single atoms with single photons has been at the heart of quantum physics since Max Planck's idea of a quantized energy exchange between light and matter and Albert Einstein's conclusion that a light beam consists of a stream of particles. This was the origin of the "first quantum revolution", which was the development of today's quantum theory, mainly by Schrödinger, Heisenberg and Dirac. In spite of rapid theoretical progress, however, the study of individual quantum systems has long remained elusive, leading in 1952 to the famous conclusion of Erwin Schrödinger that "[...] we never experiment with just *one* electron or atom [...]. In thought-experiments, we sometimes assume that we do; this invariably entails ridiculous consequences [...]" [1]. Things changed with the invention and demonstration of radio-frequency traps for electrons and ions, e.g. in the groups of Wolfgang Paul, Hans Dehmelt, Peter Toschek and David Wineland. Since then, the study of individual quantum systems has proceeded rapidly and the steady increase in experimental control has led to the dream of a "second quantum revolution" [2], which means the development of novel technologies that are based on quantum physics and provide functionality beyond any classical device. The possible applications range from quantum-enhanced precision measurements [3] to communication with unbreakable encryption [4], and from the simulation of complex many-body systems [5] to a fundamental enhancement of computability [6].

In the past years, the feasibility of these applications has been studied in various physical systems. For their pioneering work with trapped ions [7, 8] and microwave photons [9, 10], the Nobel prize in physics was awarded to David Wineland and Serge Haroche in 2012. Similar quantum control has meanwhile been achieved with neutral atoms [11, 12], optical photons [13], single spins in quantum dots and other solid-state host materials [14–16], superconducting circuits [17, 18], and even micromechanical oscillators [19]. Each of these systems has its own advantages, but also severe drawbacks which prevent to scale current proof-of-concept experiments to

A. Reiserer, *A Controlled Phase Gate Between a Single Atom and an Optical Photon*, Springer Theses,
DOI 10.1007/978-3-319-26548-3_1

a large number of particles, which is required to exploit the full potential of quantum technology. The main open challenge is that the individual particles forming a quantum system have to be largely decoupled from the environment to prevent decoherence, while at the same time a controllable long-range interaction is required for up-scaling. It was realized early on that a hybrid system of light and matter qubits [20] could tackle this problem: Single matter qubits can be well-isolated from the environment, while single photons can be used to connect them over large distances. Prime examples of the power of this approach are the proposed quantum repeater schemes [21, 22] that enable long-distance quantum communication over lossy channels, and distributed quantum information processing [16, 23] that might allow for universal scalable quantum computation [6].

Towards this end, an efficient or even deterministic interaction of single photons with single matter qubits is required. However, due to the small interaction cross section in free space, this goal is hard to achieve. A possible solution is the use of atomic ensembles consisting of a large number of particles to store and process single qubits in a material system. Here, proposals for long-distance communication [24] and information processing based on dipole blockade [25] exist and remarkable progress has been made [26].

The work presented in this thesis follows a different approach and instead makes use of the strong interaction between single photons and a single atom trapped in a resonator. The first experiments to develop this field called cavity quantum electrodynamics (CQED) have been carried out with Rydberg atoms in microwave resonators, with the realization of the one-atom maser [27] in the group of Herbert Walther being a first landmark. Later, in the group of Serge Haroche, the nondestructive detection of microwave photons employing their interaction with a beam of single atoms [28] even allowed repeated quantum-non-demolition (QND) measurements [29, 30]. The same group also demonstrated a quantum gate between a single microwave photon and single passing atoms [31]. Entering the regime of strong coupling to superconducting qubits as "artificial atoms" in 2004 [32] has triggered the development of a rapidly growing field, circuit quantum electrodynamics, which explores the great potential of CQED for quantum information processing [18]. However, the rapid decoherence of the qubits still poses a problem towards this goal. In addition, the use of microwave photons is restricted to a cryogenic environment, hampering the transmission of quantum states over large distances.

These problems can be avoided by using single trapped atoms, which can exhibit coherence times of many seconds [33]. In addition, the atomic state can be strongly coupled to optical photons in a high-finesse optical resonator. The first pioneering experiments in this direction have been carried out in the groups of Jeff Kimble and Gerhard Rempe, e.g. the observation of conditional phase shifts [34] or the controlled generation of single photons [35]. Especially the "vacuum stimulated Raman adiabatic passage" (STIRAP) technique [36–38] has been the basis for several landmark experiments: First, the generation of single photons with a steady increase in control and efficiency [35, 39–46]. Subsequently, the creation of atom-photon entangled states and the transfer of the atomic state onto the polarization of a single photon [47, 48].

The first experiments that we carried out in the course of this thesis also used the vSTIRAP technique. First, we extended the previously demonstrated state transfer between light and a single trapped atom [49] to polarization qubits, which allowed us to implement a single-atom quantum memory [50]. We then used this memory to store photons generated by another, similar setup, thereby creating a maximally entangled state between two atoms in remote laboratories [51]. This forms the elementary building block of a coherent quantum network [52] that consists of single atoms trapped in optical resonators. Subsequently, we demonstrated that the two setups can generate indistinguishable photons. This facilitated the use of an all-optical Bell-state measurement to implement an efficient teleportation protocol for the heralded state-transfer between remote single-atom quantum memories [53]. The details of these experiments can be found in the theses of my colleagues Specht [54] and Nölleke [55].

The experiments mentioned above already demonstrate the great potential of atom-cavity systems with respect to quantum information applications. The full exploration of this potential, however, was hindered by the imperfect localization and the remaining motion of the trapped atoms. Therefore, a three-dimensional optical lattice with high trap frequencies was implemented in the course of this thesis. In combination with two Raman laser beams, this allows for full control over the quantum state of a single atom trapped in a cavity, including atomic position, motion, internal state and atom-photon coupling strength [56]. The implemented techniques and the experimental results are described in detail in Chap. 2. The newly gained control allowed to reach the regime of strong coupling in our overcoupled resonator and thus to realize a controlled phase gate between a single atom and an optical photon. To this end, a novel interaction mechanism was experimentally implemented in this thesis, which had been proposed in 2004 by Duan and Kimble [57]. This mechanism is based on light reflection from a resonant cavity. It is briefly introduced in Sect. 1.2 and described in detail in Sect. 4.1. The mechnism is remarkably robust with respect to experimental imperfections and should be applicable in many different physical systems that reach the strong-coupling regime of CQED (defined in the following Sect. 1.2).

1.2 A Deterministic Interaction Mechanism Based on CQED

Cavity quantum electrodynamics (CQED) has been an active field of research during the past decades, and a detailed overview can be found in several textbooks, e.g. in [58–61]. Therefore, this section only gives a brief summary of the principles which are most relevant to understand the atom-photon interaction mechanism presented in this thesis.

The basic physical situation considered in CQED is a single emitter, e.g. a two-level atom, which is located in a resonator that supports one optical mode, in resonance with the atomic transition at a frequency ω. The atom exchanges energy with the electromagnetic field at a rate $2g$, where g is called the coupling strength. It is determined by the electric dipole matrix element μ_{12} and the electric field E of a photon in the resonator: $g = \frac{|\mu_{12}E|}{\hbar} = \sqrt{\frac{\mu_{12}^2\omega}{2\epsilon_0\hbar V}}$. Here, V is the volume of the resonator mode, $\hbar$ the reduced Planck constant and ϵ_0 the permittivity of free space. For a given atomic transition, the coupling strength thus only depends on V. Therefore, the use of a cavity with a small mode volume is essential in many experiments.

In addition to the coupling strength g, there are two additional parameters that set a characteristic timescale for the dynamics of the system: First, the rate κ at which the electric field of the cavity mode decays. Second, the rate γ at which the atomic dipole decays. This can be either due to atomic decay to other levels which are not part of the two-level approximation (this can be avoided using a closed transition), or due to emission of a photon at the resonant frequency, but into free space rather than into the cavity mode. When the cavity mode covers a large fraction $\Delta\Omega$ of the solid angle, the number of free-space decay modes and thus the atomic dipole decay rate can be substantially reduced. This effect, however, is usually negligible in Fabry-Perot resonators, and thus $\gamma = \gamma_0(1 - \frac{\Delta\Omega}{4\pi}) \simeq \gamma_0$, where γ_0 is the rate of free-space atomic decay.

The ratio of the three rates (g, κ, γ) determines the dynamics of the atom-cavity system. When the coupling strength g is larger than γ, atomic decay into the cavity is strongly enhanced compared to decay into free space. This is called the "Purcell-regime" and is already useful for many applications, such as photon generation. The deterministic atom-photon interaction mechanism presented in this thesis however requires operation in the regime of "strong coupling", where the atom-cavity coupling is the highest rate in the system, $g \gg (\gamma, \kappa)$. Only in this regime the reversible atom-photon coupling is faster than all irreversible processes, and perturbation theory is no longer applicable. Instead, the energy of the new eigenstates of the coupled system, also called "dressed states", can be calculated according to the model of Jaynes and Cummings [62]: $E_n^\pm = \left(n + \frac{1}{2}\right)\hbar\omega \pm \sqrt{n}\hbar g$.

The ground state of the coupled system, equivalent to the atom being in the ground state and no photon in the cavity, remains unaffected. However, the energies of the first excited states $E_1^+ = \hbar(\frac{3}{2}\omega + g)$ and $E_1^- = \hbar(\frac{3}{2}\omega - g)$ are split by $2\hbar g$. This has an important consequence: Photons that impinge onto the system resonant with the empty cavity cannot enter it any more. Instead, they are reflected, since there is no eigenstate of the coupled system with an energy that corresponds to the photonic frequency. This property is the basis for the controlled phase gate mechanism [57] implemented in this thesis. It is based on photon reflection from a resonant, single-sided cavity. While a detailed treatment will be given in Sect. 4.1, the rest of this section provides an intuitive introduction to the underlying process.

Consider a three-level atom in a lossless cavity that consists of a perfectly reflecting mirror (see Fig. 1.1, left blue disk) and a coupling mirror (right blue disk) which has a small transmission. A photon, resonant with both the cavity and the atomic transition from the ground state $|2\rangle$ to the excited state $|3\rangle$, is impinging onto and reflected from the setup (red arrows). When the atom is in the ground state $|1\rangle$ (Fig. 1.1a),

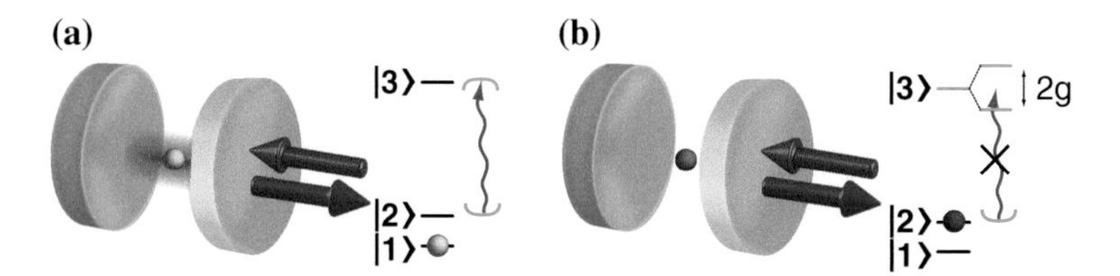

Fig. 1.1 Principle of the atom-photon interaction mechanism [63]. A resonant photon (*red arrows*) impinges onto and is reflected from an optical resonator (*blue mirrors*). **a** If the atom is in state $|1\rangle$ (*yellow*), its transitions are far detuned and the photon can enter the resonator before it is reflected. **b** If the atom is in $|2\rangle$ (*green*), strong coupling prevents the photon from entering. Instead, it is reflected directly. In this process, the combined atom-photon state experiences a conditional phase shift of π.

any transition is far detuned. Therefore, the photon can enter the cavity, as the light-field leaking out of the resonator interferes destructively with the direct reflection at the first mirror. In the reflection process, the combined atom-photon state thus experiences a phase shift of π. Now consider the case when the atom is in state $|2\rangle$ (Fig. 1.1b), strongly coupled to the resonator. Here, the photon cannot enter the cavity due to the normal-mode splitting explained above. Therefore, it is reflected without a phase shift.

The effect of the reflection process is thus a conditional phase shift of the combined atom-photon state. This controlled phase gate is implemented in this thesis. It is used to nondestructively detect the presence of a single photon (in Chap. 5) and to implement an atom-photon controlled phase gate (in Chap. 6). However, these experiments have required an unprecedented degree of control over a CQED system. The basic experimental setup to achieve this goal is not described in this work, as it has meanwhile been the topic of several doctoral theses: The design of the vacuum chamber and the atom loading and cooling procedure can be found in the thesis of Nußmann [64]; subsequent modifications, including a replacement of the cavity mirrors, are described in the theses of Specht [54] and Nölleke [55]. This thesis will instead focus on the experimental components and techniques that have not been described previously.

The outline of this thesis is as follows: In Chap. 2, the techniques to deterministically position, tightly confine and ground-state cool a single atom at a maximum of the intra-cavity field are explained, which are indispensable for the presented interaction mechanism. To exploit its full potential also requires excellent control over the internal state of the trapped atom, as well as deterministic state initialization and readout. The techniques implemented to this end are described in Chap. 3. Before turning to the main experiments, a characterization of the controlled phase gate mechanism is presented in Chap. 4. Subsequently, it is applied to nondestructively detect optical photons [63] in Chap. 5. Then, the great potential of the mechanism for quantum communication and distributed quantum computation is demonstrated by implementing a deterministic quantum gate between an optical photon and a single trapped atom [65] in Chap. 6. Finally, the results of this thesis are summarized in Chap. 7 and a short outlook to future experiments is presented.

References

1. E. Schrödinger, Are there quantum jumps? Part II. Br. J. Philos. Sci. **3**(11), 233–242 (1952). ISSN: 0007-0882. http://www.jstor.org/stable/685266
2. J.P. Dowling, G.J. Milburn Quantum technology: the second quantum revolution. Philos. Trans. R. Soc. Lond. Ser. A Math. Phys. Eng. Sci. **361**(1809), 1655–1674 (2003) ISSN: 1364-503X, 1471-2962. doi:10.1098/rsta.2003.1227. http://rsta.royalsocietypublishing.org/content/361/1809/1655
3. V. Giovannetti, S. Lloyd, L. Maccone, Quantum-enhanced measurements: beating the standard quantum limit. Science **306**(5700), 1330–1336 (2004). ISSN: 0036-8075, 1095-9203. doi:10.1126/science.1104149. http://www.sciencemag.org/content/306/5700/1330
4. N. Gisin, R. Thew, Quantum communication. Nat. Photonics **1**(3), 165–171 (2007). 00356, ISSN: 1749-4885. doi:10.1038/nphoton.2007.22. http://www.nature.com/nphoton/journal/v1/n3/abs/nphoton.2007.22.html
5. I. Buluta, F. Nori, Quantum simulators. Science **326**(5949), 108–111 (2009). ISSN: 0036-8075, 1095-9203. doi:10.1126/science.1177838. http://www.sciencemag.org/content/326/5949/108
6. T.D. Ladd et al., Quantum computers. Nature **464**(7285), 45–53 (2010). 00859, ISSN: 0028-0836. doi:10.1038/nature08812. http://dx.doi.org/10.1038/nature08812
7. D. Leibfried et al., Quantum dynamics of single trapped ions. Rev. Mod. Phys. **75**(1), 281 (2003). doi:10.1103/RevModPhys.75.281. http://link.aps.org/doi/10.1103/RevModPhys.75.281
8. D.J. Wineland, Nobel lecture: superposition, entanglement, and raising Schrödinger's cat. Rev. Mod. Phys. **85**(3), 1103–1114 (2013). doi:10.1103/RevModPhys.85.1103. http://link.aps.org/doi/10.1103/RevModPhys.85.1103
9. J.M. Raimond, M. Brune, S. Haroche, Manipulating quantum entanglement with atoms and photons in a cavity. Rev. Mod. Phys. **73**(3), 565–582 (2001). doi:10.1103/RevModPhys.73.565. http://link.aps.org/doi/10.1103/RevModPhys.73.565
10. S. Haroche, Nobel lecture: controlling photons in a box and exploring the quantum to classical boundary. Rev. Mod. Phys. **85**(3), 1083–1102 (2013). doi:10.1103/RevModPhys.85.1083. http://link.aps.org/doi/10.1103/RevModPhys.85.1083
11. R. Grimm, M. Weidemüller, Y.B. Ovchinnikov, Optical Dipole Traps for Neutral Atoms, in *Advances In Atomic, Molecular, and Optical Physics*, vol 42 (Academic Press, 2000), pp. 95–170. ISBN: 978-0-12-003842-8. http://www.sciencedirect.com/science/article/pii/S1049250X0860186X
12. I. Bloch, Quantum coherence and entanglement with ultracold atoms in optical lattices. Nature **453**(7198), 1016–1022 (2008). 00194. ISSN: 0028-0836. doi:10.1038/nature07126. http://www.nature.com/nature/journal/v453/n7198/full/nature07126.html
13. J.L. O'Brien, A. Furusawa, J. Vuckovic, Photonic quantum technologies. Nat. Photonics **3**(12), 687–695 (2009). ISSN: 1749-4885. doi:10.1038/nphoton.2009.229. http://dx.doi.org/10.1038/nphoton.2009.229
14. R. Hanson et al., Coherent dynamics of a single spin interacting with an adjustable spin bath. Science **320**(5874), 352–355 (2008). 00233 PMID: 18339902. ISSN: 0036-8075, 1095-9203. doi:10.1126/science.1155400. http://www.sciencemag.org/content/320/5874/352
15. F.A. Zwanenburg et al., Silicon quantum electronics. Rev. Mod. Phys. **85**(3), 961–1019 (2013). 00049. doi:10.1103/RevModPhys.85.961. http://link.aps.org/doi/10.1103/RevModPhys.85.961
16. D.D. Awschalom et al., Quantum spintronics: engineering and manipulating atom-like spins in semiconductors. Science **339**(6124) 1174–1179 (2013). 00072, ISSN: 0036-8075, 1095-9203. doi:10.1126/science.1231364. http://www.sciencemag.org/content/339/6124/1174
17. J.Q. You, F. Nori, Atomic physics and quantum optics using superconducting circuits. Nature **474**(7353), 589–597 (2011). ISSN: 0028-0836. doi:10.1038/nature10122. http://www.nature.com/nature/journal/v474/n7353/full/nature10122.html

18. M.H. Devoret, R.J. Schoelkopf, Superconducting circuits for quantum information: an outlook. Science **339**(6124), 1169–1174 (2013). ISSN: 0036-8075, 1095-9203. doi:10.1126/science.1231930. http://www.sciencemag.org/content/339/6124/1169
19. M. Aspelmeyer, T.J. Kippenberg, F. Marquardt, Cavity optomechanics. Rev. Mod. Phys. **86**(4), 1391–1452 (2014). 00002, doi:10.1103/RevModPhys.86.1391. http://link.aps.org/doi/10.1103/RevModPhys.86.1391
20. C. Monroe, Quantum information processing with atoms and photons. Nature **416**(6877), 238–246 (2002). ISSN: 0028-0836. doi:10.1038/416238a. http://www.nature.com/nature/journal/v416/n6877/abs/416238a.html
21. H.-J. Briegel et al., Quantum repeaters: the role of imperfect local operations in quantum communication. Phys. Rev. Lett. **81**(26), 5932–5935 (1998). 01541, doi:10.1103/PhysRevLett.81.5932. http://link.aps.org/doi/10.1103/PhysRevLett.81.5932
22. S.J. van Enk, J.I. Cirac, P. Zoller, Photonic channels for quantum communication. Science **279**(5348), 205–208 (1998). ISSN: 0036-8075, 1095-9203. doi:10.1126/science.279.5348.205. http://www.sciencemag.org/content/279/5348/205
23. C. Monroe, J. Kim, Scaling the ion trap quantum processor. Science **339**(6124), 1164–1169 (2013). ISSN: 0036-8075, 1095-9203. doi:10.1126/science.1231298. http://www.sciencemag.org/content/339/6124/1164
24. L.-M. Duan et al., Long-distance quantum communication with atomic ensembles and linear optics. Nature **414**(6862), 413–418 (2001). ISSN: 0028-0836. doi:10.1038/35106500. http://www.nature.com/nature/journal/v414/n6862/abs/414413a0.html
25. M.D. Lukin et al., Dipole blockade and quantum information processing in mesoscopic atomic ensembles. Phys. Rev. Lett. **87**(3), 037901 (2001). doi:10.1103/PhysRevLett.87.037901. http://link.aps.org/doi/10.1103/PhysRevLett.87.037901
26. K. Hammerer, A.S. Sørensen, E.S. Polzik, Quantum interface between light and atomic ensembles. Rev. Mod. Phys. **82**(2), 1041–1093 (2010). doi:10.1103/RevModPhys.82.1041. http://link.aps.org/doi/10.1103/RevModPhys.82.1041
27. D. Meschede, H. Walther, G. Müller, One-atom maser. Phys. Rev. Lett. **54**(6), 551–554 (1985). 01169, doi:10.1103/PhysRevLett.54.551. http://link.aps.org/doi/10.1103/PhysRevLett.54.551
28. G. Nogues et al., Seeing a single photon without destroying it. Nature **400**(6741), 239–242 (1999). ISSN: 0028-0836. doi:10.1038/22275. http://www.nature.com/nature/journal/v400/n6741/abs/400239a0.html
29. C. Guerlin et al., Progressive field-state collapse and quantum non-demolition photon counting. Nature **448**(7156), 889–893 (2007). 00280, ISSN: 0028-0836. doi:10.1038/nature06057. http://dx.doi.org/10.1038/nature06057
30. S. Gleyzes et al., Quantum jumps of light recording the birth and death of a photon in a cavity. Nature **446**(7133), 297–300 (2007). 00337, ISSN: 0028-0836. doi:10.1038/nature05589. http://www.nature.com/nature/journal/v446/n7133/abs/nature05589.html
31. A. Rauschenbeutel et al., Coherent operation of a tunable quantum phase gate in cavity QED. Phys. Rev. Lett. **83**(24), 5166–5169 (1999). 00547, doi:10.1103/PhysRevLett.83.5166. http://link.aps.org/doi/10.1103/PhysRevLett.83.5166
32. A. Wallraff et al., Strong coupling of a single photon to a superconducting qubit using circuit quantum electrodynamics. Nature **431**(7005), 162–167 (2004). ISSN: 0028-0836. doi:10.1038/nature02851. http://www.nature.com/nature/journal/v431/n7005/abs/nature02851.html
33. C. Deutsch et al., Spin self-rephasing and very long coherence times in a trapped atomic ensemble. Phys. Rev. Lett. **105**(2), 020401 (2010). doi:10.1103/PhysRevLett.105.020401. http://link.aps.org/doi/10.1103/PhysRevLett.105.020401
34. Q.A. Turchette et al., Measurement of conditional phase shifts for quantum logic. Phys. Rev. Lett. **75**(25), 4710–4713 (1995). doi:10.1103/PhysRevLett.75.4710. http://link.aps.org/doi/10.1103/PhysRevLett.75.4710
35. M. Hennrich et al., Vacuum-stimulated raman scattering based on adiabatic passage in a high-finesse optical cavity. Phys. Rev. Lett. **85**(23), 4872–4875 (2000). doi:10.1103/PhysRevLett.85.4872. http://link.aps.org/doi/10.1103/PhysRevLett.85.4872

36. C.K. Law, J.H. Eberly, Arbitrary control of a quantum electromagnetic field. Phys. Rev. Lett. **76**)(7), 1055–1058 (1996). 00389, doi:10.1103/PhysRevLett.76.1055. http://link.aps.org/doi/10.1103/PhysRevLett.76.1055
37. C.K. Law, H.J. Kimble, Deterministic generation of a bit-stream of single-photon pulses. J. Mod. Opt. **44**(11-12), 2067–2074 (1997). ISSN: 0950-0340. doi:10.1080/09500349708231869. http://www.tandfonline.com/doi/abs/10.1080/09500349708231869
38. A. Kuhn et al., Controlled generation of single photons from a strongly coupled atom-cavity system. Appl. Phys. B **69**(5-6), 373–377 (1999). ISSN: 0946-2171, 1432-0649. doi:10.1007/s003400050822. http://link.springer.com/article/10.1007/s003400050822
39. A. Kuhn, M. Hennrich, G. Rempe, Deterministic single-photon source for distributed quantum networking. Phys. Rev. Lett. **89**(6), 067901 (2002). doi:10.1103/PhysRevLett.89.067901. http://link.aps.org/doi/10.1103/PhysRevLett.89.067901
40. T. Legero et al., Quantum beat of two single photons. Phys. Rev. Lett. **93**(7), 070503 (2004). doi:10.1103/PhysRevLett.93.070503. http://link.aps.org/doi/10.1103/PhysRevLett.93.070503
41. J. McKeever et al., Deterministic generation of single photons from one atom trapped in a cavity. Science **303**(5666), 1992–1994 (2004). ISSN: 0036-8075, 1095-9203. doi:10.1126/science.1095232. http://www.sciencemag.org/content/303/5666/1992
42. M. Keller et al., Continuous generation of single photons with controlled waveform in an ion-trap cavity system. Nature **431**(7012), 1075–1078 (2004). 00472, ISSN: 0028-0836. doi:10.1038/nature02961. http://www.nature.com/nature/journal/v431/n7012/abs/nature02961.html
43. M. Hijlkema et al., A single-photon server with just one atom. Nat. Phys. **3**(4), 253–255 (2007). 00227, ISSN: 1745 2473. doi:10.1038/nphys569. http://dx.doi.org/10.1038/nphys569
44. T. Wilk et al., Polarization-controlled single photons. Phys. Rev. Lett. **98**(6), 063601 (2007). 00112, doi:10.1103/PhysRevLett.98.063601. http://link.aps.org/doi/10.1103/PhysRevLett.98.063601
45. G.S Vasilev, D. Ljunggren, A. Kuhn, Single photons made-to-measure. New J. Phys. **12**(6), 063024 (2010). 00027, ISSN: 1367-2630. doi:10.1088/1367-2630/12/6/063024. http://iopscience.iop.org/1367-2630/12/6/063024
46. M. Mücke et al., Generation of single photons from an atom-cavity system. Phys. Rev. A **87**(6), 063805 (2013). 00002, doi:10.1103/PhysRevA.87.063805. http://link.aps.org/doi/10.1103/PhysRevA.87.063805
47. T. Wilk et al., Single-atom single-photon quantum interface. Science **317**(5837), 488–490 (2007). doi:10.1126/science.1143835. http://www.sciencemag.org/cgi/content/abstract/317/5837/488
48. B. Weber et al., Photon-photon entanglement with a single trapped atom. Phys. Rev. Lett. **102**(3), 030501 (2009). 00082, doi:10.1103/PhysRevLett.102030501. http://link.aps.org/doi/10.1103/PhysRevLett.102.030501
49. A.D. Boozer et al., Reversible state transfer between light and a single trapped atom. Phys. Rev. Lett. **98**(19), 193601 (2007). doi:10.1103/PhysRevLett.98.193601. http://link.aps.org/doi/10.1103/PhysRevLett.98.193601
50. H.P. Specht et al., A single-atom quantum memory. Nature **473**(7346), 190–193 (2011). 00128, ISSN: 0028-0836. doi:10.1038/nature09997. http://dx.doi.org/10.1038/nature09997
51. S. Ritter et al., An elementary quantum network of single atoms in optical cavities. Nature **484**(7393), 195–200 (2012). ISSN: 0028-0836. doi:10.1038/nature11023. http://www.nature.com/nature/journal/v484/n7393/abs/nature11023.html
52. J.I. Cirac et al., Quantum state transfer and entanglement distribution among distant nodes in a quantum network. Phys. Rev. Lett. **78**(16), 3221–3224 (1997). doi:10.1103/PhysRevLett.78.3221. http://link.aps.org/doi/10.1103/PhysRevLett.78.3221
53. C. Nölleke et al., Efficient teleportation between remote single-atom quantum memories. Phys. Rev. Lett. **110**(14), 140403 (2013). 00034, doi:10.1103/PhysRevLett.110.140403. http://link.aps.org/doi/10.1103/PhysRevLett.110.140403
54. H. Specht, Einzelatom-Quantenspeicher für Polarisations-Qubits. Ph.D. Thesis. Technische Universität München (2010). http://mediatum.ub.tum.de/node?id=1002627

55. C. Nölleke, Quantum state transfer between remote single atoms. 00000. Ph.D. Thesis. Technische Universität München, (2013). http://mediatum.ub.tum.de/node?id=1145613
56. A. Reiserer et al., Ground-state cooling of a single atom at the center of an optical cavity. Phys. Rev. Lett. **110**(22), 223003 (2013). doi:10.1103/PhysRevLett.110.223003. http://link.aps.org/doi/10.1103/PhysRevLett.110.223003
57. L.-M. Duan, H.J. Kimble, Scalable photonic quantum computation through cavity- assisted interactions. Phys. Rev. Lett. **92**(12), 127902 (2004). 00427, doi:10.1103/PhysRevLett.92.127902. http://link.aps.org/doi/10.1103/PhysRevLett.92.127902
58. M. Fox, *Quantum Optics: An Introduction*. (Oxford University Press, Apr. 2006), 00288. ISBN: 978-0-19-152425-7
59. D.F. Walls, G. Gerard, J. Milburn, *Quantum Optics*. (Springer, 2008). ISBN: 3-540-28574-1
60. G.S Agarwal, *Quantum Optics*. (Cambridge University Press, 2013). ISBN: 978-1-107-00640-9
61. S. Haroche, J.-M Raimond, *Exploring the Quantum: Atoms, Cavities, and Photons*. (OUP Oxford, Apr. 2013). ISBN: 978-0-19-968031-3
62. E.T. Jaynes, F.W. Cummings, Comparison of quantum and semiclassical radiation theories with application to the beam maser. Proc. IEEE **51**(1), 89–109 (1963). ISSN: 0018-9219. doi:10.1109/PROC.1963.1664
63. A. Reiserer, S. Ritter, G. Rempe, Nondestructive detection of an optical photon. Science **342**(6164), 1349–1351 (2013). doi:10.1126/science.1246164. http://www.sciencemag.org/content/342/6164/1349
64. S. Nußmann, Kühlen und Positionieren eines Atoms in einem optischen Resonator. 00000. Ph.D. Thesis. Technische Universität München (2006). http://mediatum.ub.tum.de/node?id=603119
65. A. Reiserer et al., A quantum gate between a flying optical photon and a single trapped atom. Nature **508**(7495), 237–240 (2014). 00010, ISSN: 0028-0836. doi:10.1038/nature13177. http://www.nature.com/nature/journal/v508/n7495/full/nature13177.html

Chapter 2
Controlling the Position and Motion of a Single Atom in an Optical Cavity

2.1 Trapping Atoms in a Cavity

CQED with single trapped atoms has a long history. The first experiments in the optical domain employed hot atomic beams, with a stepwise reduction in the number of atoms in the cavity [1], which eventually led to the observation of a normal-mode splitting with an average atom number of only one [2]. In this setting, a first measurement of the phase shift that a single atom can imprint onto a transmitted faint laser beam was demonstrated [3]. However, atom-light interaction was limited to very short times ($\sim\mu s$) in these experiments. To improve this, the techniques of laser cooling and trapping [4] were introduced to CQED. First steps along these lines were taken by releasing cold atoms from a magneto-optical trap (MOT) such that they fall through a high-finesse optical cavity [5, 6]. Interaction times were further increased with an atomic fountain with the cavity at the turning point of the atoms [7, 8]. The only way to further improve was then to employ an atom trap within the cavity.

In principle, there are two commonly used techniques for single-atom trapping: First, electrical trapping of charged atoms—usually cations [9]. Second, optical trapping of neutral atoms in far detuned laser fields [10, 11]. In cavity QED experiments, the first approach—the use of ions—has been restricted to very long cavities [12, 13] due to the technical difficulty that the dielectric mirrors of optical resonators tend to disturb the electric trapping potential. Therefore, the strong-coupling regime has not been reached with trapped ions until today, albeit there is some progress due to the development of fiber-based Fabry-Perot cavities [14] which—due to their smaller size—allow to trap an ion in a resonator of sufficiently small mode volume to fulfill the condition $g > \gamma$ [15].

In contrast, the strong-coupling regime has been achieved in many experiments with trapped neutral atoms. Remarkably, it has early been demonstrated in two independent experiments in the strong coupling regime [16, 17] that even the force of single optical photons can suffice to trap an atom when operating at a resonant

A. Reiserer, *A Controlled Phase Gate Between a Single Atom and an Optical Photon*, Springer Theses,
DOI 10.1007/978-3-319-26548-3_2

electronic transition. In all experiments that employ several atomic levels, however, a state-insensitive trap is required. In this respect, the first milestone was the achievement of a 28 ms atom storage time in a far-off-resonance dipole trap (FORT) [18], implemented in a standing-wave configuration along the cavity axis. The working principle of such a trap is explained in Sect. 2.1.1. To increase the storage time for single atoms, efficient cooling mechanisms are required. A brief summary of the various previously employed techniques is given in Sect. 2.1.2, before turning to the trap implemented in this thesis in Sect. 2.2.

2.1.1 Optical Dipole Traps

To date, all experiments that trap single atoms in a cavity in the strong-coupling regime employ optical dipole traps, which have been established as a standard technique in atomic physics, summarized in several review articles (e.g. [10, 11]) and standard textbooks (e.g. [4]). While a brief introduction to the most relevant parameters is given in the following, a more detailed discussion of the geometry and parameters of the trap implemented in this thesis will be presented in Sect. 2.2.

The origin of the confining force in an optical trap is the electromagnetic field of an intense, off-resonant laser beam. In any polarizable particle, this leads to an induced dipole moment which then interacts dispersively with the gradient of the laser field. For a two-level atom with an excited state that decays to the ground state at a rate Γ while emitting photons at a frequency ω_0, the interaction potential takes the form [10]:

$$U_{dip}\left(\vec{r}\right) = -\frac{3\pi c^2}{2\omega_0^3}\left(\frac{\Gamma}{\omega_0 - \omega} + \frac{\Gamma}{\omega_0 + \omega}\right) I\left(\vec{r}\right) \tag{2.1}$$

Here, $I\left(\vec{r}\right)$ denotes the spatially dependent laser intensity, ω the laser frequency, and c the speed of light. In a two-level system, the energy of the ground state is lowered if $\omega < \omega_0$, i.e. if the laser is detuned towards the red side of the electromagnetic spectrum. The energy of the excited state is increased by the same amount. When considering alkaline atoms rather than the idealized two-level approximation, the sum over all possible transitions has to be evaluated to calculate the overall optical potential. For ^{87}Rb, the wavelength dependence of the atomic level shifts is shown in Fig. 2.1.

The spatial structure of the ground-state potential is given by the intensity distribution of the used trapping laser. A common configuration is the use of a retro-reflected laser beam, which is also called a one-dimensional optical lattice. Along the direction of the laser beam (z), the formation of a standing-wave pattern gives rise to a modulated potential with a period of $\frac{\lambda}{2}$. Along the perpendicular axes (x, y), the potential is given by the Gaussian mode profile of the laser beam with a waist of w_0. Thus, the resulting overall potential of depth U_0 takes the form:

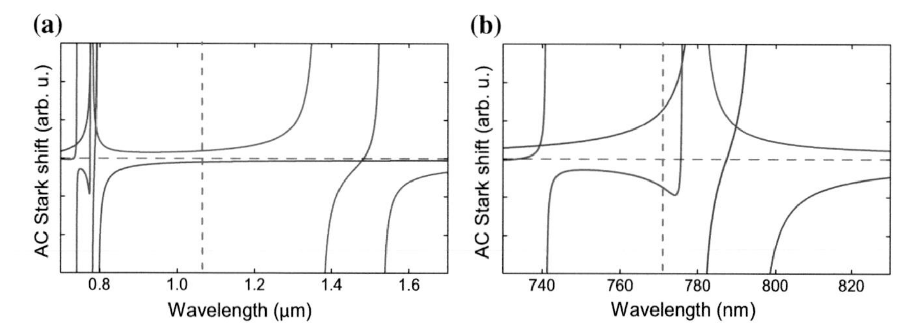

Fig. 2.1 **a** and **b** Stark shift of the atomic states $5S_{1/2}$ (*blue*) and $5P_{3/2}$ (*red*), depending on the wavelength of the trapping laser. The *dashed green lines* indicate the wavelengths of the optical lattice used in the experimental setup, described in detail in Sect. 2.2

$$U_{SW}(\vec{r}) = U_0 \cos^2\left(2\pi\frac{z}{\lambda}\right)\exp\left(-\frac{x^2+y^2}{2w_0^2}\right) \tag{2.2}$$

When several non-interfering laser beams are employed (i.e. using orthogonal polarizations), the individual potentials simply add up. The same holds when the beams have different frequencies and thus the interference pattern shifts its position on a timescale that is fast compared to the atomic motion. Close to its bottom, the trap is well approximated by a harmonic potential. In this case, the vibrational energy states are equidistant with a separation of $h\nu_{trap}$, where ν_{trap} is called the trap frequency. Along the standing-wave axis, it is given by: $\nu_{trap} = \frac{\hbar}{\lambda}\sqrt{\frac{2U_0}{m}}$.

When the trap frequency is much larger than the recoil frequency that corresponds to the momentum of a single resonant photon, the motional state of the atom does not change in most absorption and emission events. This regime is called the Lamb-Dicke regime and is very favorable for efficient laser cooling to the ground state of the potential. For Rubidium atoms, the recoil frequency is $2\pi \cdot 4$ kHz, which means that deep potentials with trap frequencies on the order of 10–100 kHz are required. This can be achieved with high laser intensities, however at the price of increased heating and ground-state decoherence, since the rate at which the atom scatters trap light is given by:

$$R_{scat}(\vec{r}) = \frac{3\pi c^2}{2\hbar\omega_0^3}\left(\frac{\omega}{\omega_0}\right)^3\left(\frac{\Gamma}{\omega_0-\omega}+\frac{\Gamma}{\omega_0+\omega}\right)^2 I(\vec{r}) \tag{2.3}$$

A possible solution is the choice of a larger detuning $\Delta \equiv \omega_0 - \omega$, as the trap depth scales inversely proportional to Δ, the scattering rate however inversely proportional to Δ^2. This is the reason why most experiments with optical dipole traps operate at a large detuning of several THz.

2.1.2 Techniques to Control the Position and Motion of Atoms in a Cavity

In cavity QED with trapped atoms, the demonstration of new physical effects has often required an increase in control over the external degrees of freedom of the atoms. Therefore, the attempt to accurately localize an atom at a well-defined position in the cavity field and to eliminate its motion has a very long history. After the first successful trapping experiments [18], there have been numerous approaches to improve the localization and extend the storage times.

An early approach was to gain information about the atomic position and motion from the temporal modulation of a resonant laser beam transmitted through the cavity [16, 17]. Subsequent feedback onto the trap allowed to increase the atomic storage time [19]. With further improvement of the experimental setup and the electronics [20] and a higher output-coupling efficiency of the cavity, cooling of the atomic motion and storage times exceeding 1 s have been demonstrated [21].

The first QED experiments with second-long atom trapping times [22], however, used a different cooling method, namely a combination of red-detuned Doppler cooling (on a closed transition) and blue-detuned Sisyphus cooling (on another transition). In these experiments, Cs atoms were trapped in a magic-wavelength [11] FORT in a standing-wave configuration along the cavity axis. In search of faster cooling rates and lower temperatures, a mechanism has been proposed [23, 24] that directly makes use of the coupling to a cavity. The first implementation [25] of this cavity-cooling method used Rb atoms, also confined in an intra-cavity dipole trap.

The improved atom trapping in both of the above mentioned experimental setups led to the simultaneous achievement of another milestone in cavity QED—the observation of a normal mode spectrum with a single trapped atom [26, 27]. Achieving even longer atom trapping times was hampered by large parametric heating rates due to fluctuations of the used intra-cavity traps. This problem was avoided in a novel setup that used a cavity-independent FORT. With a combination of vacuum-stimulated and Sisyphus cooling, trapping times of 17 s were observed [28]. After replacement of the cavity and subsequent bake-out, this value was increased in the course of this thesis to more than 1 min on average [29], which is most likely limited by collisions with the background gas.

In addition to the increased storage times, the cavity-independent trap has another advantage: it allows to change the position of the atom along the beam axis with sub-micron precision [30] by shifting the standing-wave pattern of the trapping laser. This technique was later used in two different setups to deterministically transfer an atom to the cavity center [31, 32]. To this end, the exact number and position of atoms trapped in a standing-wave FORT has been determined with fluorescence imaging. Subsequently, the atoms were shifted to the center of a cavity using an optical conveyor belt [33]. In the experimental setup used in this thesis, this approach has been extended further. A high-numerical-aperture objective has been implemented which collects the light scattered by atoms trapped within the cavity mode [34]. Using the images of an electron-multiplying CCD camera then facilitates feedback onto the position of the standing-wave trap [29], which allows us to localize the trapped atoms at a well-defined position within the cavity.

The above mentioned cooling mechanisms all lead to the observation of long storage times. However, it is also important to cool the atoms to a low temperature to avoid motion-induced dephasing, fluctuations in the atom-cavity coupling strength and a time-varying AC Stark shift of the atomic levels in case the trap is not a magic one [11]. In principle, cavity-cooling is a promising scheme for this purpose. With a long cavity of high finesse, even cooling below the recoil limit has been demonstrated with an atomic ensemble [35]. Unfortunately, cavity cooling requires the operation at a specific detuning, which prevents all experiments that operate at a different detuning or on resonance. The same holds for the recently proposed [36] and observed [37, 38] cavity EIT cooling. Therefore, the use of another technique is highly desirable. Along these lines, Raman sideband cooling—a very successful technique known from free-space experiments with trapped ions [9, 39, 40] and neutral atoms [41–46]—has been demonstrated in one dimension. Along the axis of the used intra-cavity dipole trap, the motional ground state has been achieved with a probability of 95 % [47].

To summarize, the control over motion and position of single atoms in cavity QED has been steadily improved over the past decades. In this thesis, this culminates in the achievement of full control, as explained in detail in Sect. 2.4. The decisive step towards this goal was the implementation of a three-dimensional optical lattice with high trap frequencies in all directions, described in Sect. 2.2. In addition, the current apparatus builds on many of the above mentioned techniques, especially on the use of a cavity-independent dipole trap with intra-cavity Sisyphus cooling [28] and Raman sideband cooling [47], a blue-detuned intra-cavity trap [48], imaging [34] and active atom positioning [30, 33]. As a side remark, these techniques and the increasing experimental complexity have raised the demands on the stability of the setup. Therefore, literally every part of the laser system, the optics setup and the frequency stabilization electronics has been replaced and rebuilt in the course of this thesis.

2.2 Implementation of a 3D Optical Lattice in a Cavity

Previous approaches to atom trapping in a cavity used only one retro-reflected laser beam, as described in 2.1.2. This only provides subwavelength confinement in the direction of the standing-wave pattern, i.e. along the laser beam axis, which leads to several experimental drawbacks. First, even at low temperatures, the residual atomic motion is usually on a scale that is comparable to the wavelength of the used lasers. This prevents all experiments that require phase-stable illumination of the atom. When the trap axis does not coincide with that of the resonator, the atomic motion also poses the problem of strongly fluctuating atom-cavity coupling strength. In addition, the trap frequencies in the orthogonal directions are comparably small, typically on the order of a few kHz. This can be the cause of reduced atom trapping times due to parametric heating, caused by acoustic vibrations of the optics setup in this frequency regime.

In this thesis, the aforementioned obstacles have been eliminated by implementing a 3D optical lattice within the cavity. The trapping geometry has been designed such that the atoms can be deterministically located at the maximum of the intra-cavity field to achieve the maximum possible coupling strength. This is especially challenging along the axis of the resonator, where the cavity mode exhibits a standing-wave structure and the atoms thus have to be localized with subwavelength precision. Therefore, one of the axes of the optical lattice has been chosen to coincide with the cavity axis. In this configuration, the standing-wave structure of the cavity and that of the lattice always exhibit a fixed spatial relation. In addition, the reflection or the transmission of the trapping laser can be used to derive an error signal that allows to stabilize the cavity resonance frequency with the Pound-Drever-Hall technique [49].

Unfortunately, due to the AC Stark shift induced by the trap light, this geometry can also have two severe drawbacks: First, fluctuations of the cavity length, e.g. due to uncompensated mechanical vibrations, will affect the intensity of the trap laser and thus lead to fluctuating atomic transition frequencies. Second, circular polarization components of the trap can arise from cavity birefringence. This leads to differential shifts of the atomic ground states and thus to decoherence of superposition states when the intensity fluctuates or the atoms move in the trap. Both problems can be minimized by using a blue-detuned lattice beam along the cavity axis. In this case, cold atoms are trapped at a node of the field, such that they experience low absolute AC Stark shifts even at high trap intensities. With respect to the absolute value of the trap laser detuning, one has to consider the beating pattern between the blue-detuned trap and the intra-cavity field at the atomic resonance frequency, which can be seen in Fig. 2.2a. When the relative detuning corresponds to an odd number of free spectral ranges, a maximum of the intra-cavity field (red) will coincide at the center of the resonator with a minimum of the trapping potential (blue), such that the atom (black dot) is trapped at a position with maximum coupling strength. In the experiments described in this thesis, the blue-detuned trap was operated at a wavelength of 771 nm, which leads to a beating period of about 32 μm, which poses only moderate requirements on the accuracy of the atom positioning.

In contrast to the trap along the cavity axis, the lattice beams along the orthogonal directions can be implemented with a high degree of experimental flexibility. In the existing setup, the use of a horizontally and a vertically oriented trap seemed to be favorable in terms of optical access. We chose to use a blue-detuned (771 nm) and a red detuned (1064 nm) laser beam. The resulting configuration is illustrated in Fig. 2.2b (not to scale). Before the laser setup is explained in more detail, the reasons for the used wavelength configuration are presented in the following.

There are two major advantages of using a blue-detuned trap along the vertical axis: First, at low temperatures the scattering-induced heating rate of a blue trap is smaller than that of a red trap at the same detuning [10]. Second, the atoms are trapped at an antinode of the standing wave, which means that the AC Stark shift is close to zero, independent of the exact intensity of the trap light. Therefore, the

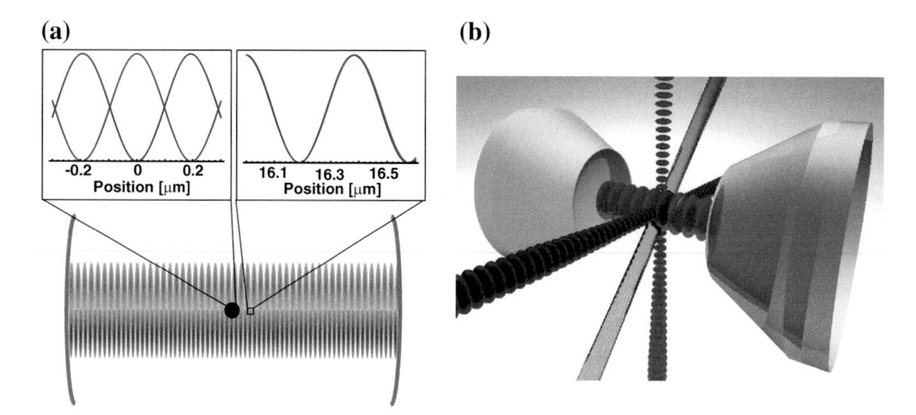

Fig. 2.2 **a** Spatial structure of the intra-cavity dipole trap (not to scale). The atom (*black dot*) is trapped in a blue-detuned standing-wave trap (*blue*) which has a different wavelength than the cavity mode (*red*). Thus, the atom-cavity coupling depends on the atomic position along the resonator axis. This is indicated in the plots above. When the atom is trapped at the center of the resonator (0 μm, *left*), it experiences the maximum coupling strength, while the coupling is very small when the atom is trapped at a position 16 μm away from the center (*right*), such that the cavity mode and the trap light largely overlap. **b** Sketch of the setup geometry (not to scale). The atoms are trapped at the center of a resonator that is made of two highly-reflective spherical mirrors (*light blue*), which are coned to provide better optical access. The atoms are trapped in a three-dimensional optical lattice which consists of three orthogonal standing-wave fields, one red-detuned (along the horizontal axis) and two blue-detuned. The atoms can be addressed with laser beams of different frequencies that impinge either along the cavity axis or from the side (*orange*), oriented perpendicular to the cavity axis and at an angle of 45° with respect to the other trapping beams

frequencies of the atomic transitions remain constant, even if the spatial position of the laser beams that form the 3D optical lattice drifts over time.

For the horizontal axis, however, we chose to employ a red-detuned laser beam operating at 1064 nm. This also has several advantages: First, the attractive potential of the resulting standing-wave trap allows to efficiently load atoms into the lattice [50]. Second, this wavelength has proven to facilitate efficient cooling of the trapped atoms using a Sisyphus-like mechanism [28]. Third, the atoms are trapped at a position of maximum intensity. When the trap light is linearly polarized along the quantization axis, this leads to a Zeeman-state dependent level splitting of the excited state (described in more detail in Sect. 3.1, see Fig. 3.1), which has proven useful for efficient optical pumping [51] and for the implementation of the quantum gate described in Chap. 6 without an interferometrically stable optics setup. However, the use of a red detuned trap also has a caveat: It is very important that the trap light does not exhibit any circular polarization components, as they would lead to considerable differential AC Stark shifts of the ground-state Zeeman levels. This could lead to a severe reduction in coherence time if the atom is moving in the trap or if the trap intensity fluctuates.

In the following, the laser setup of both traps is explained in more detail, starting with the red-detuned one. The previously existing setup of this trap [52] had used electro-optical modulators for the required trap switching. These modulators, however, turned out to be very unstable with respect to thermal fluctuations and have therefore been replaced with AOMs. In addition, optical fibers have been introduced into the beam path to improve the spatial profile of the trapping beam and to reduce the sensitivity of the setup with respect to temperature fluctuations. In order to ensure that the trap light does not exhibit any circular polarization components, a Semrock polarizing bandpass filter is used. This yields a measured polarization extinction ratio of $10^6 : 1$, which is an improvement by three orders of magnitude compared to the previous setup. Apart from these modifications, the geometry is described in the thesis of Stephan Nußmann [50].

In the following, the setup of the vertical trap is explained. The lattice beam originates from the same laser as that of the intra-cavity trap. To prevent any interference effects, the polarization of both beams is set orthogonal. In addition, they are detuned by several hundred MHz with respect to each other, such that a possibly remaining interference pattern is averaged out as it changes on a timescale that is fast compared to the atomic motion. The laser beam impinges to the vacuum chamber from the bottom, meaning that all optics can be placed close to the surface of the optical table, where a stable mounting is guaranteed. The beam is focused to the atoms with an achromatic doublet lens of 2.5 cm focal length, leading to a diffraction-limited spot size of about 10 μm (FWHM) at the focal position. After passing through the vacuum chamber, the beam is collimated by the same objective that is also used to image the atoms. This objective was designed by A. Kochanke and provides a nearly diffraction-limited resolution at the two design wavelengths 780 and 1064 nm. The 771 nm trapping light is retro-reflected after separation from the atomic fluorescence with several bandpass filters.

The fluorescence light is focused onto an electron-multiplying CCD camera, which allows to record the atom distribution with up to five images per second. Figure 2.3 shows two typical images, one with the intracavity trap switched off (left), and one in the 3D-lattice configuration (right). The tighter confinement of the atom can be directly seen, as its size is clearly reduced in the vertical direction in the image, which corresponds to the cavity axis.

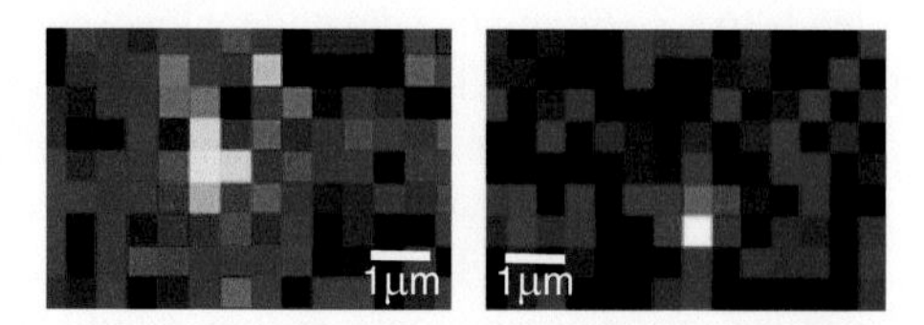

Fig. 2.3 Fluorescence images of single trapped atoms with the intra-cavity trap switched off (*left*) or on (*right*), respectively. The presence of the confining potential can be seen directly in the images, as it reduces the atomic extent along the vertical axis in the images, which corresponds to the cavity axis. The color scale is linear, normalized to the highest intensity value of each image

2.3 Deterministic Localization of a Single Atom at the Center of the Resonator

In the following section, the techniques to localize an atom at a predefined position in the resonator are explained. The contents of this section have been published in [53]: *Ground-State Cooling of a Single Atom at the Center of an Optical Cavity*. A. Reiserer, C. Nölleke, S. Ritter, G. Rempe. Physical Review Letters **110**, 223003 (2013).

In our setup, each experimental run starts with the preparation of a cloud of ^{87}Rb atoms in a magneto-optical trap (MOT). To transfer the atoms to the optical resonator, a running-wave dipole trap is then used. When the atoms arrive in the cavity, they are first loaded into the red-detuned 1D optical lattice using the procedure described in [30, 50]. Subsequently, cooling light is applied in a counter-propagating geometry from the side of the resonator (orange laser beam in Fig. 2.2). The cooling light is 30 MHz red-detuned with respect to the $F = 2 \leftrightarrow F' = 3$ transition of the D_2 line and has orthogonal linear polarizations, which leads to cooling of the atom in all three dimensions using a Sisyphus-type mechanism [28, 50]. A high numerical aperture objective is used to collect light that is scattered by the atom, which allows to image the atoms and to determine their number and position using an algorithm that evaluates the recorded intensity pattern. The loading procedure is repeated until a single atom is detected in the images [54]. Subsequently, the standing-wave pattern is shifted along the beam, which allows to deterministically transfer the atom to an arbitrary position within the cavity mode [30]. Subsequently, the lattice beams described in Sect. 2.2 are switched on, such that the atom is tightly confined along all directions.

In the following, experimental control of the coupling strength is demonstrated by loading the atoms at different positions with respect to the cavity field. We measure the transmission of a weak probe laser pulse, which is resonant with the empty cavity and the Stark-shifted atomic transition from $|F, m_F\rangle = |2, 2\rangle$ to $|3, 3\rangle$. Here, F denotes the atomic hyperfine state and m_F its projection onto the quantization axis, which coincides with the cavity axis. Depending on the coupling strength, the transmission is suppressed, as explained in Sect. 1.2. Shifting the atom along the axis of the red detuned trap therefore gives a Gaussian dependence (red dots in Fig. 2.4a), due to the Gaussian radial profile of the cavity mode. When shifting the red-detuned dipole trap along the cavity axis (using a piezo mirror), a beating between the sinusoidal variation of the effective coupling strength g and the standing-wave trap along the cavity axis is expected, see Fig. 2.2a and [12, 13, 55]. This is shown in Fig. 2.4a (black squares).

We observe a sinusoidal modulation of the transmission. The deviation from the ideally expected oscillation with the same period but steeper slopes is caused by a position-dependent optical pumping efficiency and temperature, which leads to averaging effects in coupling strength and Stark shift. Due to the loading procedure, the

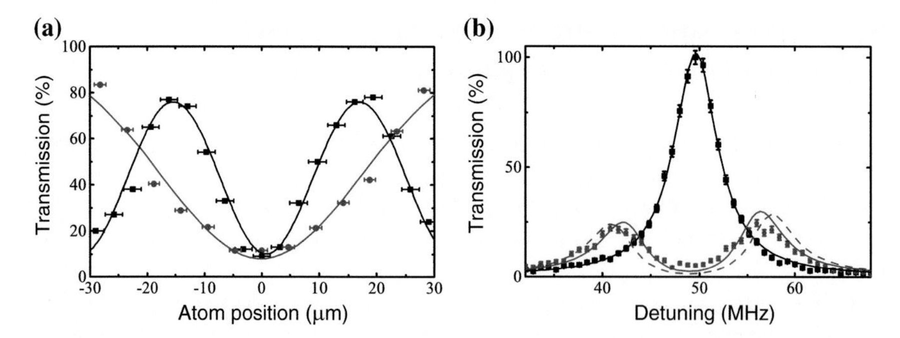

Fig. 2.4 **a** Transmission through the cavity when the position of a single, coupled atom is scanned along the cavity axis (*black squares*) and along an orthogonal axis (*red dots*). The transmission is strongly suppressed when the atom is located at a maximum of the intracavity field. The solid Gaussian (*red*) and sine (*black*) fit curves are a guide for the eye. **b** Normal-mode spectroscopy of the atom-cavity system with the atom trapped in the 3D optical lattice. The transmission of the cavity is a Lorentzian curve when the atom is not coupled (*black squares* and *black fit curve*), while a resonant atom leads to a normal-mode splitting (*red dots* and *solid red fit curve*). The slight asymmetry is caused by a small residual detuning between atom and cavity. The error bars are statistical. The dashed curve shows the spectrum expected for $g_0/2\pi = 8\,\text{MHz}$, the value calculated from our cavity parameters (from [53])

initial distribution of the atoms in the lattice is determined by their initial temperature and the beam waist of the red-detuned dipole trap. We determine the width of this distribution from the fluorescence images. This gives the error bars in Fig. 2.4a. On the length scale of the positioning error, the transmission is nearly constant. We can thus deterministically localize a single atom at the maximum of the resonator field, where the atom-cavity coupling is strongest.

The absolute strength of this coupling is determined by recording the normal-mode spectrum of the atom-cavity system [2]. To this end, the frequency of the probe laser is scanned while the frequency of the cavity is kept fixed. To record the spectrum of the empty cavity, the atom is first pumped to $F = 1$ such that it is not coupled to the resonator. Thus, the transmission is a Lorentzian curve with a full width at half maximum of 5.5 MHz (Fig. 2.4b, black squares). When the atom is prepared in the $|2, 2\rangle$ state, we observe a normal-mode splitting (red dots). The separation of the two peaks is twice the atom-cavity coupling constant g. To determine this value, we fit the normal modes with a theory curve (solid red line) with g and the atomic detuning as the only free parameters. From this fit, we find $g/2\pi = (6.7 \pm 0.1)\,\text{MHz}$, close to the theoretical value of $g_0/2\pi = 8\,\text{MHz}$ (dashed red line). This again proves that we are able to accurately localize the atom at the center of the cavity field and that the system is in the single-atom strong coupling regime of CQED.

2.4 Sideband Spectroscopy and Ground-State Cooling

In Sect. 2.3, excellent control over the position of a single atom with respect to the cavity mode has been demonstrated. This section will present techniques to control its motion. Again, the contents of this section have partially been published in [53].

In the implemented optical lattice (see Sect. 2.2), the atom is tightly confined in all three dimensions to a spatial extent that is small compared to the wavelength of the employed optical transitions (spatial extent of the ground-state wave function $\lesssim$15 nm). This situation is called the Lamb-Dicke regime, in which very powerful techniques to analyze and control the atomic motion exist. These techniques have been pioneered in free space experiments with single trapped ions [9, 40] and ensembles of neutral atoms [10, 56, 57]. Recently, they have been applied to single neutral atoms trapped in optical tweezers [45, 46] and—in this thesis—in a three-dimensional optical lattice. The basic idea is that two Raman laser beams allow one to drive transitions between the different motional states of the atom. While the used setup will be described in more detail in Sect. 3.2, this section will focus on its application to atom cooling.

In our experiment, the atom is trapped in a three-dimensional optical lattice at a temperature that is small compared to the trap depth. Thus, the confining potential can be well approximated by a three-dimensional harmonic potential, as explained in more detail in Sect. 2.1.1. In this situation, the atomic motion is quantized with equally spaced energy levels with $E = (n + \frac{1}{2}) \cdot h\nu_{trap}$ [10]. A pair of Raman lasers is employed, which exhibit a relative detuning that equals the hyperfine splitting of 6.8 GHz and a common detuning of 0.3 THz with respect to the D_1 line at 795 nm. Because of this large detuning, the Raman beams lead to an effective coupling of the two hyperfine ground states without populating the excited state. The linewidth of this coupling can be much smaller than the natural linewidth of the D_1 transition. This allows one to drive atomic transitions that increase or lower the vibrational quantum number n, as schematically depicted in Fig. 2.5a. The coupling strength of these transitions is given by the following formulas [9]:

$$\begin{aligned} \Omega_{n \to n-1} &= \Omega_{n \to n} \eta \sqrt{n} \\ \Omega_{n \to n+1} &= \Omega_{n \to n} \eta \sqrt{n+1} \end{aligned} \tag{2.4}$$

Here, $\Omega_{n \to n}$ denotes the Rabi frequency of the carrier transition, which depends on the geometry, polarization and intensity of the used laser beams. $\Omega_{n \to n-1}$ ($\Omega_{n \to n+1}$) is the Rabi frequency of a transition on the red (blue) sideband, which lowers (increases) the vibrational quantum number by one. $\eta = \sqrt{\frac{\hbar}{4\pi m \nu_{trap,x}}} \Delta k_x$ is called the Lamb-Dicke parameter, which depends on the atomic mass m and Δk_x, the wavevector difference of the two Raman laser beams, projected onto the axis of motion $\vec{x}$. In our experiment, $\eta \simeq 0.1$. From Eq. (2.4), one can directly see that the coupling strength of the sideband transitions depends on the vibrational quantum number n. When R

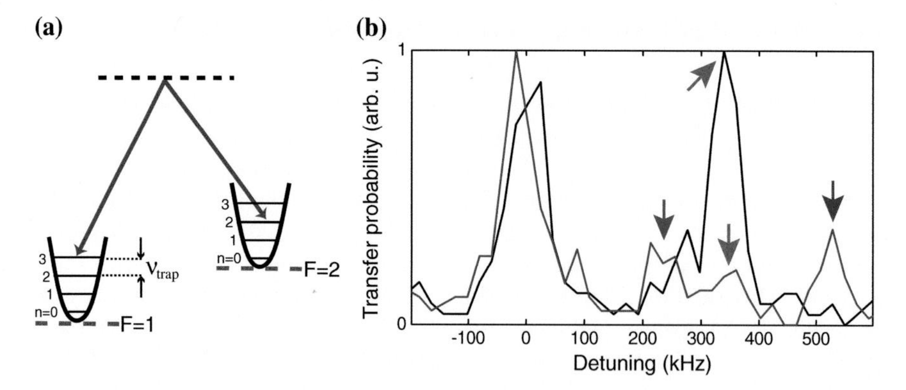

Fig. 2.5 **a** Coupling of different motional states with Raman lasers. The atom is confined in a harmonic potential, which is schematically depicted for the two ground states $F = 1$ and $F = 2$. With the right detuning of the Raman laser beams (*red arrows*), states with different vibrational quantum number can be coupled. **b** Amplitude of the Raman carrier and the blue sideband. When the atom is trapped at the center of the cavity (*black line*), only the central sideband (*green arrow*) that corresponds to the trap along the cavity axis is resolved. When the atomic position is shifted by $\sim 5\,\mu m$ (*red line*) by shifting the red-detuned trap, all three sidebands are visible with comparable height (colored arrows; *blue* vertical axis; *green* cavity axis; *red* horizontal axis)

denotes the ratio of the transition probabilities on the red and blue sidebands, the average vibrational quantum number $\bar{n}$ of a thermal state is given by [9]:

$$\bar{n} = \frac{R}{1 - R} \tag{2.5}$$

Thus, it is possible to determine the atomic temperature when the sidebands can be resolved in a spectroscopic measurement. To experimentally obtain a sideband spectrum, the atom is optically pumped to the $F = 1$ hyperfine state. Subsequently, the Raman lasers are applied for a certain time period, typically $200\,\mu s$. In order to measure the population transfer to $F = 2$, cavity-based hyperfine state detection is employed [54], which will be described in more detail in Sect. 3.1.

To analyze and control the atomic motion, we first used a geometry where one of the beams was applied along the cavity, while the other beam was applied from the side, forming an angle of 45° with respect to the other axes of the optical lattice, compare Fig. 2.2b. The beams were orthogonally polarized, thus driving transitions for any of the atomic Zeeman states. However, it turned out that in this configuration, the coupling strength of the individual sidebands strongly depends on the atomic position within the standing-wave Raman beam, similar to what has been observed in [47]. The reason is that the standing-wave Raman field has a fixed phase of α with respect to the intracavity trap, which yields $\Omega_{n\rightarrow n\pm1}^{\text{intracavity}} = \Omega_{n\rightarrow n\pm1}\sin(2\alpha)$ [47]. This effect is not present for the orthogonal axes, where the Raman beams do not exhibit a standing-wave structure. Due to the change in α, the relative amplitude of the intracavity sideband with respect to those of the other axes changes when the atom is trapped at different positions along the resonator axis. This effect can be seen in the sideband spectrum of Fig. 2.5b. When the atom is trapped at the center of the cavity (black line), the sidebands of the orthogonal traps are much smaller than that of the intracavity trap (green arrow), while they can be clearly resolved (blue and red arrow) when the atomic position is shifted by $\sim 5\,\mu\text{m}$ (red line) along the cavity axis.

In order to cool the atom in all three dimensions at the center of the cavity, however, it is highly advantageous to have equal coupling strengths on all three sidebands. Therefore, an additional Raman laser beam is employed, which is counter-propagating to the beam that is applied from the side. Its polarization is set orthogonal to both other beams to prevent interference effects. It drives transitions on all sidebands except that along the cavity axis. Thus, the relative amplitude of the other sidebands can be increased with the power in the additional Raman beam. With this, the amplitude of all three sidebands is in the following set to a comparable height, and the atomic temperature after intra-cavity Sisyphus cooling is investigated.

The green line in Fig. 2.6a shows an obtained sideband spectrum, where zero detuning means a frequency difference that corresponds to the hyperfine transition frequency. The large peak at the center of the spectrum is the saturated carrier transition. At negative detunings, the red sidebands can be seen, corresponding to transitions that lower the vibrational state of the atom by one quantum. The three peaks at positive detunings correspond to the blue sideband for each of the three lattice axes: the red-detuned dipole trap (at 0.5 MHz) and the blue-detuned traps along the vertical axis (0.4 MHz) and along the cavity axis (0.3 MHz). The peaks can be identified unambiguously by successively changing the intensity of one of the lattice beams and then recording the sideband spectrum (not shown). The central sideband peak

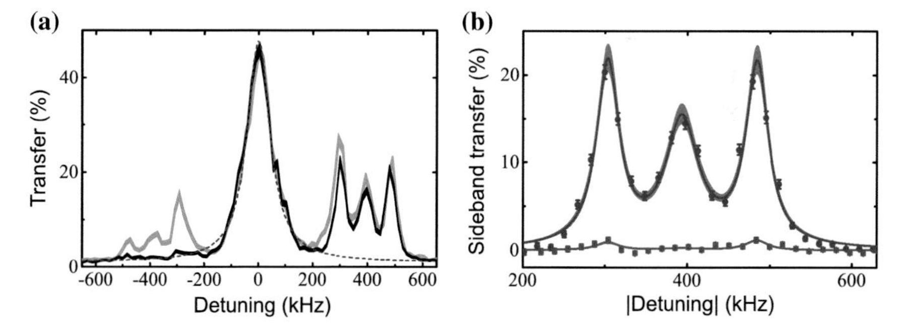

Fig. 2.6 **a** Sideband spectrum after intra-cavity Sisyphus (*green*) and after sideband cooling (*black*). The statistical standard error of the data is given by the thickness of the lines. The *three peaks* at positive detunings correspond to a transition on the blue sideband for each axis of the 3D lattice potential (*right* to *left*: $\hat{x}$, $\hat{y}$, $\hat{z}$ axis). The carrier peak at the center (*dashed blue* Lorentzian fit curve) is saturated. Transitions on the red sideband (negative detunings) are still observed after Sisyphus cooling (*green*) but nearly vanish after 5 ms of sideband cooling (*black*). **b** Transfer probability on the red (*red squares*) and blue (*blue dots*) sideband after Raman sideband cooling. The solid curves are numerical fits of the sum of three Lorentzian curves, with the *shaded areas* indicating the 66 % confidence interval. The atomic temperature after sideband cooling is determined from these fits (from [53])

is lower and broader than the other two in the depicted long-term measurement. On shorter timescales, three peaks of the same height are observed, but with fluctuating position of the central peak. This is caused by long-term drifts in beam pointing, as at the time of the depicted measurement a lattice beam with a much tighter focus was used along the vertical axis due to the limited laser power available. Meanwhile, this laser has been replaced by one that gives a much higher output power, and the spot size has been doubled to reduce the trap-frequency fluctuations.

Applying (2.5) to a fit of the green curve in Fig. 2.6a gives $\bar{n}_{\{x,y,z\}} = \{0.19(5), 0.4(1), 1.0(2)\}$. Here, x corresponds to the axis of the red-detuned trap, y to the vertical and z to the cavity axis. This demonstrates that the intra-cavity Sisyphus cooling mechanism already leads to temperatures well below the Doppler limit [28, 50] ($\bar{n}_{\mathrm{D}} \approx 6$–$10$ for our trap frequencies). To further reduce the atomic temperature, pulsed Raman sideband cooling is used. To this end, the atom is prepared in $F = 1$ and the Raman beams are applied for 5 ms with frequency components that drive transitions on all three red sidebands. During this interval, a $\approx$10 ns long repump pulse is applied on the $F = 2 \leftrightarrow F' = 1$ transition every 200 ns in order to bring any transferred population back to $F = 1$ where the cooling cycle can start again. To determine the effect of the sideband cooling, the following measurement cycle is performed: First, a 4.4 ms long interval of intra-cavity Sisyphus cooling is applied on the closed transition. Then, the transfer probability at a certain Raman detuning is recorded. Subsequently, sideband cooling is applied and the transfer probability at the same detuning is measured again. This measurement sequence is repeated at different frequencies to record a spectrum immediately before (green in Fig. 2.6a) and after (black) sideband cooling. The red sidebands vanish almost completely, which indicates that the atom is cooled close to the ground state.

To determine the mean occupation number $\bar{n}$, a Lorentzian curve is fit to the carrier (blue dashed line in Fig. 2.6a) and subtracted from the data (Fig. 2.6b). Subsequently, the sum of three Lorentzian curves is fit to the blue sidebands (blue curve) to determine the width and frequency of the three peaks as well as their respective amplitudes. Using the same frequencies and widths for the red sidebands, their amplitude is determined, again from a least-squares fit to the data (red curve). This gives $\bar{n}_{\{x,y,z\}} = \{0.04(1), 0.02(1), 0.06(1)\}$. Assuming a thermal distribution, this means that the atom is cooled to the 1D ground state with a probability of $\{0.96(1), 0.98(1), 0.95(1)\}$ and to the 3D ground state with a probability of (89 ± 2) %.

One of the main advantages of cooling to low temperatures is that the atomic transition frequencies are expected to be constant, even without the use of a trap at a magic wavelength [11]. To investigate whether this is the case in our experimental setup, we perform a measurement of the AC Stark shift. To this end, the atom is prepared in $F = 1$ and a laser resonant with the transition to $|1', \pm 1\rangle$ is irradiated. The frequency of this transfer laser is scanned and the transfer probability to $F = 2$ is measured, conditioned on atom trapping at the center of the red-detuned dipole trap using the camera images. This gives the spectrum in Fig. 2.7. While this measurement was performed after ground-state cooling, a similar result can be obtained in the 3D lattice using only intra-cavity Sisyphus cooling (not shown), as this already leads to sufficiently low temperatures (as demonstrated above).

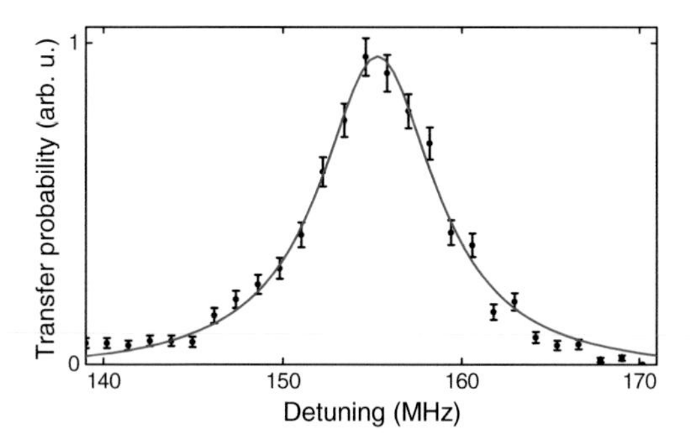

Fig. 2.7 AC Stark shift of the atomic transition after cooling the atom to the three-dimensional motional ground state. The measured linewidth (*red Lorentzian fit curve*) is 8 MHz FWHM, close to the natural linewidth of the atomic transition (6 MHz). The remaining broadening is due to atom trapping at different positions within the red-detuned dipole trap, which are not resolved in the camera images. The *error bars* denote the standard error of the mean

The observed spectrum has a Lorentzian linewidth of 8 MHz (FWHM) (red fit curve), which is slightly broader than the natural linewidth of the atomic transition (6 MHz). The remaining deviation is caused by different atom-trapping positions in the lattice, which are not resolved in the camera images, but can still lead to a slightly different intensity of the 1064 nm light. Without conditioning on the atom position, a larger linewidth of about 12 MHz is observed. Still, this value constitutes some improvement compared to previous experiments in our setup, where a typical value of 23 MHz (FWHM) was observed [52]. This improvement was a prerequisite for the observation of a clear normal-mode splitting [53] (see Sect. 2.4) and thus for the experiments presented in Chaps. 5 and 6 [58, 59].

References

1. G. Rempe et al., Optical bistability and photon statistics in cavity quantum electrodynamics. Phys. Rev. Lett. **67**(13), 1727–1730 (1991). doi:10.1103/PhysRevLett.67.1727. http://link.aps.org/doi/10.1103/PhysRevLett.67.1727
2. R.J. Thompson, G. Rempe, H.J. Kimble, Observation of normal-mode splitting for an atom in an optical cavity. Phys. Rev. Lett. **68**(8), 1132–1135 (1992). 00722, doi:10.1103/PhysRevLett.68.1132. http://link.aps.org/doi/10.1103/PhysRevLett.68.1132
3. Q.A. Turchette et al., Measurement of conditional phase shifts for quantum logic. Phys. Rev. Lett. **75**(25), 4710–4713 (1995). doi:10.1103/PhysRevLett.75.4710. http://link.aps.org/doi/10.1103/PhysRevLett.75.4710
4. J. Harold, in *Metcalf and Peter Van der Straten*. Laser Cooling And Trapping, (Springer, 1999). ISBN: 978-0-387-98728-6
5. C.J. Hood et al., Real-time cavity QED with single atoms. Phys. Rev. Lett. **80**(19), 4157–4160 (1998). doi:10.1103/PhysRevLett.80.4157. http://link.aps.org/doi/10.1103/PhysRevLett.80.4157
6. M. Hennrich et al., Vacuum-stimulated raman scattering based on adiabatic passage in a high-finesse optical cavity. Phys. Rev. Lett. **85**(23), 4872–4875 (2000). doi:10.1103/PhysRevLett.85.4872. http://link.aps.org/doi/10.1103/PhysRevLett.85.4872

7. P. Münstermann et al., Dynamics of single-atom motion observed in a high-finesse cavity. Phys. Rev. Lett. **82**(19), 3791–3794 (1999). doi:10.1103/PhysRevLett.82.3791. http://link.aps.org/doi/10.1103/PhysRevLett.82.3791
8. P. Münstermann et al., Observation of cavity-mediated long-range light forces between strongly coupled atoms. Phys. Rev. Lett. **84**(18), 4068–4071 (2000). doi:10.1103/PhysRevLett.84.4068. http://link.aps.org/doi/10.1103/PhysRevLett.84.4068
9. D. Leibfried et al., Quantum dynamics of single trapped ions. Rev. Mod. Phys. **75**(1), 281 (2003). doi:10.1103/RevModPhys.75.281. http://link.aps.org/doi/10.1103/RevModPhys.75.281
10. R. Grimm, M. Weidemüller, Y.B. Ovchinnikov, Optical dipole traps for neutral atoms. Adv. At. Mol. Opt. Phys. **42**, 95–170 (Academic Press, 2000). ISBN: 978-0-12-003842-8. http://www.sciencedirect.com/science/article/pii/S1049250X0860186X
11. A. Derevianko, H. Katori, Colloquium: physics of optical lattice clocks. Rev. Mod. Phys. **83**(2), 331–347 (2011). doi:10.1103/RevModPhys.83.331. http://link.aps.org/doi/10.1103/RevModPhys.83.331
12. G.R. Guthöhrlein et al., A single ion as a nanoscopic probe of an optical field. Nature **414**(6859), 49–51 (2001). ISSN: 0028-0836. doi:10.1038/35102129. http://www.nature.com/nature/journal/v414/n6859/abs/414049a0.html
13. A.B. Mundt et al., Coupling a single atomic quantum bit to a high finesse optical cavity. Phys. Rev. Lett. **89**(10), 103001 (2002). 00291, doi:10.1103/PhysRevLett.89.103001. http://link.aps.org/doi/10.1103/PhysRevLett.89.103001
14. D. Hunger et al., A fiber Fabry-Perot cavity with high finesse. New J. Phys. **12**(6), 065038 (2010). ISSN: 1367-2630. doi:10.1088/1367-2630/12/6/065038. http://iopscience.iop.org/1367-2630/12/6/065038
15. M. Steiner et al., Single ion coupled to an optical fiber cavity. Phys. Rev. Lett. **110**(4), 043003 (2013). 00015, doi:10.1103/PhysRevLett.110.043003. http://link.aps.org/doi/10.1103/PhysRevLett.110.043003
16. C.J. Hood et al., The atom-cavity microscope: single atoms bound in orbit by single photons. Science **287**(5457), 1447–1453 (2000). ISSN: 0036-8075, 1095-9203. doi:10.1126/science.287.5457.1447. http://www.sciencemag.org/content/287/5457/1447
17. P.W.H. Pinkse et al., Trapping an atom with single photons. Nature **404**(6776), 365–368 (2000). ISSN: 0028-0836. doi:10.1038/35006006. http://dx.doi.org/10.1038/35006006
18. J. Ye, D.W. Vernooy, H.J. Kimble, Trapping of single atoms in cavity QED. Phys. Rev. Lett. **83**(24), 4987–4990 (1999). doi:10.1103/PhysRevLett.83.4987. http://link.aps.org/doi/10.1103/PhysRevLett.83.4987
19. T. Fischer et al., Feedback on the motion of a single atom in an optical cavity. Phys. Rev. Lett. **88**(16), 163002 (2002). 00106, doi:10.1103/PhysRevLett.88.163002. http://link.aps.org/doi/10.1103/PhysRevLett.88.163002
20. A. Kubanek et al., Photon-by-photon feedback control of a single-atom trajectory. Nature **462**(7275), 898–901 (2009). ISSN: 0028-0836. doi:10.1038/nature08563. http://dx.doi.org/10.1038/nature08563
21. M. Koch et al., Feedback cooling of a single neutral atom. Phys. Rev. Lett. **105**(17), 173003 (2010). 00034, doi:10.1103/PhysRevLett.105.173003. http://link.aps.org/doi/10.1103/PhysRevLett.105.173003
22. J. McKeever et al., State-insensitive cooling and trapping of single atoms in an optical cavity. Phys. Rev. Lett. **90**(13), 133602 (2003). 00324, doi:10.1103/PhysRevLett.90.133602. http://link.aps.org/doi/10.1103/PhysRevLett.90.133602
23. P. Horak et al., Cavity-induced atom cooling in the strong coupling regime. Phys. Rev. Lett. **79**(25), 4974–4977 (1997). 00252, doi:10.1103/PhysRevLett.79.4974. http://link.aps.org/doi/10.1103/PhysRevLett.79.4974
24. V. Vuletic, S. Chu, Laser cooling of atoms, ions, or molecules by coherent scattering. Phys. Rev. Lett. **84**(17), 3787–3790 (2000). 00259, doi:10.1103/PhysRevLett.84.3787. http://link.aps.org/doi/10.1103/PhysRevLett.84.3787

25. P. Maunz et al., Cavity cooling of a single atom. Nature **428**(6978), 50–52 (2004). ISSN: 0028-0836. doi:10.1038/nature02387. http://dx.doi.org/10.1038/nature02387
26. A. Boca et al., Observation of the vacuum rabi spectrum for one trapped atom. Phys. Rev. Lett. **93**(23), 233603 (2004). doi:10.1103/PhysRevLett.93.233603. http://link.aps.org/doi/10.1103/PhysRevLett.93.233603
27. P. Maunz et al., Normal-mode spectroscopy of a single-bound-atom-cavity system. Phys. Rev. Lett. **94**(3), 033002 (2005). doi:10.1103/PhysRevLett.94.033002. http://link.aps.org/doi/10.1103/PhysRevLett.94.033002
28. S. Nußmann et al., Vacuum-stimulated cooling of single atoms in three dimensions. Nat. Phys. **1**(2), 122–125 (2005). ISSN: 1745-2473. doi:10.1038/nphys120. http://www.nature.com/nphys/journal/v1/n2/abs/nphys120.html
29. H.P. Specht et al., A single-atom quantum memory. Nature **473**(7346), 190–193 (2011). 00128, ISSN: 0028-0836. doi:10.1038/nature09997. http://dx.doi.org/10.1038/nature09997
30. S. Nußmann et al., Submicron positioning of single atoms in a microcavity. Phys. Rev. Lett. **95**(17), 173602 (2005). doi:10.1103/PhysRevLett.95.173602. http://link.aps.org/doi/10.1103/PhysRevLett.95.173602
31. K.M. Fortier et al., Deterministic loading of individual atoms to a high-finesse optical cavity. Phys. Rev. Lett. **98**(23), 233601 (2007). doi:10.1103/PhysRevLett.98.233601. http://link.aps.org/doi/10.1103/PhysRevLett.98.233601
32. M. Khudaverdyan et al., Controlled insertion and retrieval of atoms coupled to a high-finesse optical resonator. New J. Phys. **10**(7), 073023 (2008). 00053, ISSN: 1367-2630. doi:10.1088/1367-2630/10/7/073023. http://iopscience.iop.org/1367-2630/10/7/073023
33. S. Kuhr et al., Deterministic delivery of a single atom". Science **293**(5528), 278–280 (2001). ISSN: 0036-8075, 1095-9203. doi:10.1126/science.1062725. http://www.sciencemag.org/content/293/5528/278
34. B. Weber et al., Photon-photon entanglement with a single trapped atom. Phys. Rev. Lett. **102**(3), 030501 (2009). 00082, doi:10.1103/PhysRevLett.102.030501. http://link.aps.org/doi/10.1103/PhysRevLett.102.030501
35. M. Wolke et al., Cavity cooling below the recoil limit. Science **337**(6090), 75–78 (2012). 00027, ISSN: 0036-8075, 1095-9203. doi:10.1126/science.1219166. http://www.sciencemag.org/content/337/6090/75
36. M. Bienert, G. Morigi, Cavity cooling of a trapped atom using electromagnetically induced transparency. New J. Phys. **14**(2), 023002 (2012). 00016, ISSN: 1367-2630. doi:10.1088/1367-2630/14/2/023002. http://iopscience.iop.org/1367-2630/14/2/023002
37. T. Kampschulte et al., Optical control of the refractive index of a single atom. Phys. Rev. Lett. **105**(15), 153603 (2010). 00067, doi:10.1103/PhysRevLett.105.153603. http://link.aps.org/doi/10.1103/PhysRevLett.105.153603
38. T. Kampschulte et al., Electromagnetically-induced-transparency control of single-atom motion in an optical cavity. Phys. Rev. A **89**(3), 033404 (2014). 00000, doi:10.1103/PhysRevA.89.033404. http://link.aps.org/doi/10.1103/PhysRevA.89.033404
39. F. Diedrich et al., Laser cooling to the zero-point energy of motion. Phys. Rev. Lett. **62**(4), 403–406 (1989). 00693, doi:10.1103/PhysRevLett.62.403. http://link.aps.org/doi/10.1103/PhysRevLett.62.403
40. C. Monroe et al., Resolved-sideband Raman cooling of a bound atom to the 3d zero-point energy. Phys. Rev. Lett. **75**(22), 4011–4014 (1995), doi:10.1103/PhysRevLett.75.4011. http://link.aps.org/doi/10.1103/PhysRevLett.75.4011
41. H.J. Lee et al., Raman cooling of atoms in an optical dipole trap. Phys. Rev. Lett. **76**(15), 2658 (1996), doi:10.1103/PhysRevLett.76.2658. http://link.aps.org/doi/10.1103/PhysRevLett.76.2658
42. V. Vuletic et al., Degenerate Raman sideband cooling of trapped cesium atoms at very high atomic densities. Phys. Rev. Lett. **81**(26), 5768–5771 (1998). 00210. doi:10.1103/PhysRevLett.81.5768. http://link.aps.org/doi/10.1103/PhysRevLett.81.5768
43. S.E. Hamann et al., Resolved-sideband Raman cooling to the ground state of an optical lattice. Phys. Rev. Lett. **80**(19), 4149–4152 (1998). doi:10.1103/PhysRevLett.80.4149. http://link.aps.org/doi/10.1103/PhysRevLett.80.4149

44. H. Perrin et al., Sideband cooling of neutral atoms in a far-detuned optical lattice. Europhys. Lett. (EPL) **42**(4), 395–400 (1998). ISSN: 0295-5075. doi:10.1209/epl/i1998-00261-y. http://iopscience.iop.org/0295-5075/42/4/395
45. A.M. Kaufman, B.J. Lester, C.A. Regal, Cooling a single atom in an optical tweezer to its quantum ground state. Phys. Rev. X**2**(4), 041014 (2012). doi:10.1103/PhysRevX.2.041014. http://link.aps.org/doi/10.1103/PhysRevX.2.041014
46. J.D. Thompson et al., Coherence and Raman sideband cooling of a single atom in an optical tweezer. Phys. Rev. Lett. **110**(13), 133001 (2013). doi:10.1103/PhysRevLett.110.133001. http://link.aps.org/doi/10.1103/PhysRevLett.110.133001
47. A.D. Boozer et al., Cooling to the ground state of axial motion for one atom strongly coupled to an optical cavity. Phys. Rev. Lett. **97**(8), 083602 (2006). doi:10.1103/PhysRevLett.97.083602. http://link.aps.org/doi/10.1103/PhysRevLett.97.083602
48. T. Puppe et al., Trapping and observing single atoms in a blue-detuned intracavity dipole trap. Phys. Rev. Lett. **99**(1), 013002 (2007). 00073, doi:10.1103/PhysRevLett.99.013002. http://link.aps.org/doi/10.1103/PhysRevLett.99.013002
49. E.D. Black, An introduction to Pound-Drever-Hall laser frequency stabilization. Am. J. Phys. **69**(1), 79 (2001). ISSN: 00029505. doi:10.1119/1.1286663. http://link.aip.org/link/AJPIAS/v69/i1/p79/s1&Agg=doi
50. S. Nußmann, Kühlen und Positionieren eines Atoms in einem optischen Resonator". 00000. PhD thesis. Technische Universität München (2006). http://mediatum.ub.tum.de/node?id=603119
51. C. Nölleke, Quantum state transfer between remote single atoms. 00000. PhD thesis. Technische Universität München (2013). http://mediatum.ub.tum.de/node?id=1145613
52. H. Specht, Einzelatom-Quantenspeicherfür Polarisations-Qubits. PhD thesis. Technische Universität München (2010). http://mediatum.ub.tum.de/node?id=1002627
53. A. Reiserer et al., Ground-state cooling of a single atom at the center of an optical cavity. Phys. Rev. Lett. **110**(22), 223003 (2013). doi:10.1103/PhysRevLett.110.223003. http://link.aps.org/doi/10.1103/PhysRevLett.110.223003
54. J. Bochmann et al., Lossless state detection of single neutral atoms. Phys. Rev. Lett. **104**(20), 203601 (2010). 00051, doi:10.1103/PhysRevLett.104.203601. http://link.aps.org/doi/10.1103/PhysRevLett.104.203601
55. Y. Colombe et al., Strong atom-field coupling for Bose-Einstein condensates in an optical cavity on a chip. Nature **450**(7167), 272–276 (2007). ISSN: 0028-0836.doi:10.1038/nature06331. http://www.nature.com/nature/journal/v450/n7167/abs/nature06331.html
56. A.J. Kerman et al., Beyond optical molasses: 3D Raman sideband cooling of atomic cesium to high phase-space density. Phys. Rev. Lett. **84**(3), 439–442 (2000). doi:10.1103/PhysRevLett.84.439. http://link.aps.org/doi/10.1103/PhysRevLett.84.439
57. D.J. Han et al., 3D Raman sideband cooling of cesium atoms at high density. Phys. Rev. Lett. **85**(4), 724–727 (2000). 00097, doi:10.1103/PhysRevLett.85.724. http://link.aps.org/doi/10.1103/PhysRevLett.85.724
58. A. Reiserer, S. Ritter, G. Rempe, Nondestructive detection of an optical photon. Science **342**(6164), 1349–1351 (2013). doi:10.1126/science.1246164. http://www.sciencemag.org/content/342/6164/1349
59. A. Reiserer et al., A quantum gate between a flying optical photon and a single trapped atom. Nature **508**(7495), 1349–1351 (2014). 00010, ISSN: 0028-0836. doi:10.1038/nature13177. http://www.nature.com/nature/journal/v508/n7495/full/nature13177.html

Chapter 3
Measurement and Control of the Internal Atomic State

3.1 State Initialization and State Detection

The controlled phase gate mechanism that is implemented in this thesis is based on an atomic three-level system, where two of the levels are strongly coupled via the cavity, while the other level is far detuned. This model system is of course only an approximation to the real experimental situation, as the used ^{87}Rb atoms exhibit a richer level structure, see Fig. 3.1. For the atom-photon interaction mechanism, only a small fraction of the atomic levels are used, namely the states marked in black: $|1, 1\rangle$, $|2, 2\rangle$ and $|3', 3\rangle$.

The AC Stark shift of the atomic ground and excited states can be calculated by summing over all relevant atomic transitions. To obtain the values for our experiment, the shift of the transition from $|2, 2\rangle$ to $|3, 3\rangle$ has been measured to be 0.10 GHz, where 0.05 GHz stem from the shift of the ground state $|2, 2\rangle$. The values calculated for the other excited states are: $|3, 0\rangle$: 0.16 GHz, $|3, \pm 1\rangle$: 0.15 GHz, $|3, \pm 2\rangle$: 0.10 GHz, $|3, \pm 3\rangle$: 0.05 GHz. These values are in excellent agreement with the measured results of a recent publication [1].

At the beginning of each experimental run, the atom has to be prepared in a well-defined electronic state. To this end, an optical pumping sequence is employed. A repumping laser is applied from the side of the cavity, which drives the transition from $F = 1$ to $F' = 2$ to transfer the atomic population to $F = 2$. At the same time, a pumping laser is applied along the cavity axis with right-circular polarization, such that transitions from $F = 2$ to $F' = 3$ with $\Delta m_F = +1$ are driven. In this way, the atomic state ends up in $|2, 2\rangle$ after a time interval of typically 150 µs, which is determined by the applied laser intensity.

In the state $|2, 2\rangle$, the transmission of the cavity is strongly reduced due to strong coupling, as will be explained in detail in Sect. 4.1. Therefore, measuring the transmitted light intensity allows to determine whether the atom has been pumped to the desired state [2]. To this end, single photon counting modules (SPCMs) are employed. Figure 3.2a shows a histogram of the detected photon number when the

A. Reiserer, *A Controlled Phase Gate Between a Single Atom and an Optical Photon*, Springer Theses,
DOI 10.1007/978-3-319-26548-3_3

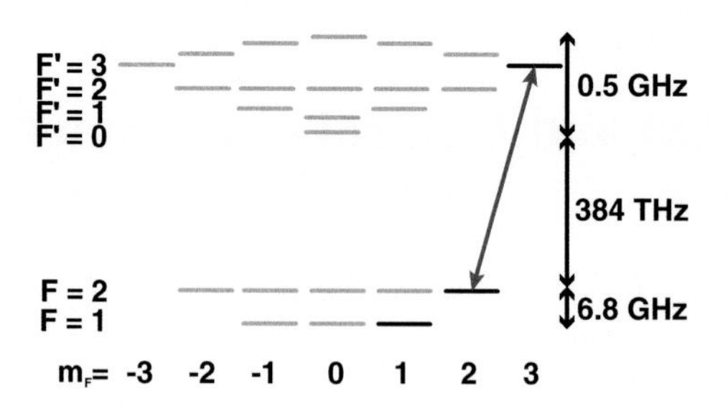

Fig. 3.1 Level Scheme of ^{87}Rb (not to scale). The transition with the highest atom-cavity coupling strength is marked in *red*. The atomic levels which are most relevant for the experiments of this thesis are shown as *black lines*. For the used trap configuration, the AC Stark shift of excited state levels in the $F' = 1$ and $F' = 3$ manifolds depends on the Zeeman state m_F (see text)

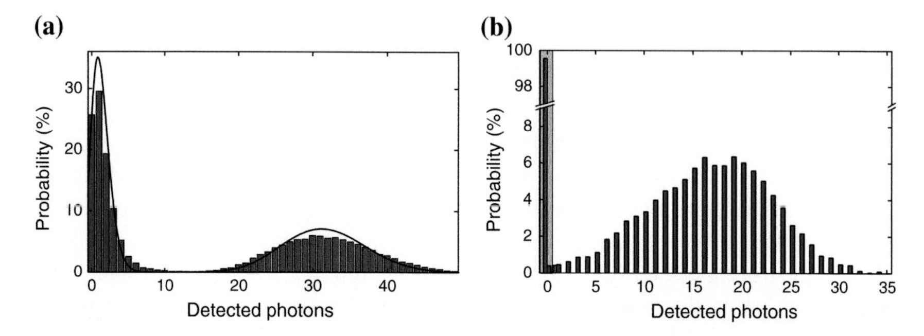

Fig. 3.2 **a** Preparation and detection of the atomic state by cavity transmission. At the beginning of the optical pumping process (*red*), the atom is in the uncoupled state $F = 1$. Measuring the transmission of the cavity thus leads to a Poissonian distribution (*black* fit curve) with an average number of about 30 detected photons within a 10 μs long interval. At the end of the pumping process, the atom is prepared in $|2, 2\rangle$ (*blue*), where the transmission of the cavity is reduced due to strong coupling. The detected transmission again follows a Poissonian distribution (*black* fit curve), but with a much smaller average photon number. As the detected distributions are clearly separated, it is also possible to detect the atomic state with a high fidelity of 98 %. **b** Detection of the atomic hyperfine state using cavity enhanced fluorescence (from [6]). A laser resonant with the cavity and the atomic transition from $|2, 2\rangle$ to $|3', 3\rangle$ is applied from the side. The atom scatters many photons into the cavity mode when it is in $F = 2$ (*blue*), whereas no photons are detected when the atom is in $F = 1$ (*red*). As the detected photon distributions are clearly separated, the technique facilitates atomic hyperfine-state detection with a fidelity of 99.65 % within 3 μs

atom is prepared in the uncoupled state $F = 1$ (red), and after optical pumping to the state $|2, 2\rangle$ (blue). In both cases, Poissonian distributions (black fit curves) are observed, with a much smaller mean photon number in the coupled case. As the two distributions are clearly separated, it is possible to detect whether the atom has been pumped to the desired state [3–5]. Using a discrimination threshold of 15 detected photons yields a value above (below) threshold in the uncoupled (coupled) state with a probability of 99.7 % (96.8 %). The latter value is limited by the probability to pump the atom to an uncoupled state in the F = 2 manifold during the state detection interval. This might be caused by imperfect circular polarization due to cavity birefringence, which means that a laser beam which is circularly polarized in front of the cavity can exhibit a slightly different polarization within the cavity mode. If the cavity frequency would be perfectly stable, this effect could be precompensated. However, no improvement could be observed in a series of measurements where the polarization of the pumping beam was systematically scanned using two waveplates.

While the above procedure allows to detect whether the atom is in state $|2, 2\rangle$ or not, another technique can be employed to detect whether it is in $F = 2$, independent of the Zeeman state. To this end, the atom is excited using a resonant laser beam and the resulting fluorescence is detected with SPCMs, as it also is a common technique in free-space ion-trap experiments [7–9]. When the atoms are trapped in a cavity, one can take advantage [4] of the Purcell effect, which directs the atomic fluorescence to the cavity mode and thus can strongly increase the photon collection efficiency. The major advantage of this technique, compared to state detection in transmission, is that the average number of detected photons is almost zero when the atom is in $F = 1$, as can be seen in the histogram in Fig. 3.2b (red). Thus, only a small number of scattered photons is sufficient to detect the state $F = 2$ (blue), which means that the detection process can be very fast, typically 3 μs. In addition, a higher state detection fidelity of 99.65 % can be achieved. Therefore, this technique is employed to detect the atomic state in the experiments described in Chaps. 5 and 6.

3.2 Raman Control of the Internal Atomic State

Breakthrough experiments in CQED have often been enabled by an increase in control over the atom-cavity system. Atom trapping and cooling has meanwhile become a standard technique. However, full control over the internal state of the atom and over its motion has not been previously achieved in a CQED setup. In this thesis, this goal has been reached by trapping the atom in a three-dimensional optical lattice and implementing a pair of Raman lasers. The use of Raman transitions has been pioneered in ion trap experiments [9–11]. A comparable degree of control over the electronic ground state of neutral atoms has been achieved much later [12, 13]. This section gives a summary of the Raman techniques implemented in this work. In Sect. 3.2.1, the basic setup of the laser system is explained. In Sect. 3.2.3, the selective addressing of individual atomic Zeeman levels is demonstrated, which is used for precise control of the atomic state, as described in Sect. 3.2.4. This control is a prerequisite for the experiments presented in Chaps. 5 and 6, as well as for numerous future studies.

3.2.1 Laser Setup

In this section, the basic parameters of our Raman laser system are summarized. It operates close to the D_1 line of Rubidium around 795 nm, such that the Raman beams can be easily separated from light at the D_2 line using optical bandpass filters. Thus, it is simultaneously possible to use SPCMs to monitor the cavity output at 780 nm and to apply one Raman beam along the cavity axis, driving a different longitudinal mode. This is in our setup indispensable for three-dimensional sideband cooling. In addition, the fluorescence images of the atom are not affected, even when high Raman powers are used. In the present setup, all Raman beams originate from the same laser. This has the advantage that it is easily possible to achieve relative phase stability without technically demanding lock electronics. The frequency difference between the two Rubidium ground states, 6.8 GHz, is bridged using a combination of electro-optical and acousto-optical modulators [14], see Fig. 3.3a.

In principle, the Raman lasers can be operated at an arbitrarily large detuning Δ from the atomic resonance. However, the goal to achieve full control over the atomic state requires the Raman transfers to be fast compared to all decoherence mechanisms. This determines a certain range of reasonable detunings: On the one hand, $|\Delta|$ should be as large as possible to avoid population of the excited states during the Raman pulses (which would lead to decoherence due to spontaneous photon emission). On the other hand, the available laser power of a few mW sets an upper bound to $|\Delta|$, as the transfer rate is inversely proportional to it. These considerations resulted in values between 0.1 THz and 0.3 THz, depending on the experimental requirements. For ground-state cooling (Sect. 2.4, [15]), the requirement to drive a cavity mode that has an intensity maximum at the center of the resonator led to

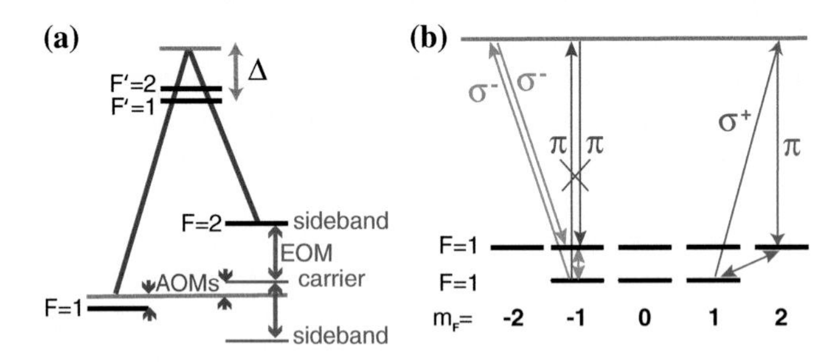

Fig. 3.3 **a** Rubidium level scheme with the employed Raman transitions (not to scale). The laser is stabilized to a frequency corresponding to the energy difference between the two *gray lines*. It is detuned by Δ from the atomic transitions on the D_1 line at 795 nm. The individual beams (*red lines*) are switched on and off and shifted in frequency using two acousto-optical modulators (AOMs). One beam additionally passes an electro-optical modulator that generates sidebands (*green arrows* and levels) at a frequency of 6.5 GHz. While two-thirds of the light in this laser beam is off-resonant with respect to any atomic transition, one third is in two-photon resonance with the other beam, i.e. the mutual detuning corresponds exactly to the frequency difference between the two hyperfine ground states $F = 1$ and $F = 2$ of ^{87}Rb (6.8 GHz). **b** Examples of Raman transitions in the Rubidium level scheme, including all hyperfine (F) and Zeeman (m_F) ground states. Due to interference, the atomic levels $|F, m_F\rangle = |1, -1\rangle$ and $|2, -1\rangle$ are not coupled when two laser beams of the same linear polarization π (*faint red arrows*) are employed. If, however, two lasers of the same circular polarizations σ^- are used (*faint green arrows*), the transition can be driven (*green double arrow*). Similarly, the levels $|1, 1\rangle$ and $|2, 2\rangle$ are coupled (*blue double arrow*) by a combination of π and σ^+ polarized light (*light blue arrows*)

$|\Delta| = 0.30\,\text{THz}$. For fast control of the atomic Zeeman state, a smaller detuning of $0.10\,\text{THz}$ (Chap. 5, [2]) or $0.15\,\text{THz}$ (Chap. 6, [6]) was employed, respectively.

3.2.2 Raman Coupling for Different Laser Polarizations

The polarization (and thus also the geometry) of the Raman beams has to be adapted to the experimental situation. To understand this, it is convenient to use the cavity axis as the quantization axis, which defines the Zeeman states in the full atomic level scheme depicted in Fig. 3.3b. A laser beam which is linearly polarized along the cavity axis drives only π transitions with $\Delta m_F = 0$. This polarization requires that the beam is impinging orthogonally from the side of the resonator. Similarly, a circularly polarized beam applied along the cavity axis would drive either σ^+ or σ^- transitions with $\Delta m_F = +1$ or $\Delta m_F = -1$, respectively. Finally, a beam that is impinging orthogonally from the side of the cavity, polarized perpendicular to its axis, would drive a coherent superposition of $\sigma^+ + \text{e}^{i\varphi}\sigma^-$. Here, the relative phase φ is determined by the direction of incidence. We define $\varphi = 0$ for a Raman beam applied along the pump laser axis (orange beam in Fig. 2.2b).

In order to determine which atomic levels are coupled by the Raman laser beams, all transitions that are driven by a certain polarization configuration have to be considered. The coupling strength can then be calculated by summing over all possible transitions, weighted with their Clebsch-Gordan coefficients (including their sign). In many configurations, this leads to destructive interference. An example is given in Fig. 3.3b, where the transition from $|1, -1\rangle$ to $|2, -1\rangle$ (green double arrow) cannot be driven by two laser beams of π polarization (faint red arrows). If, instead, two lasers of the same circular polarization, e.g. σ^-, are used (faint green arrows), the levels are coupled when the Raman beams are applied.

In the following, the transition from $|1, 1\rangle$ to $|2, 2\rangle$ (blue double arrow) is considered, since it is the most relevant transition in the context of this thesis. This transition can be driven using a combination of π and σ^+ polarized light (light blue arrows). As this requires pure σ^+ polarization, the respective beam has to drive a cavity mode. The latter, however, poses several experimental drawbacks: First, the standing-wave pattern of the Raman beam has a different period than that of the intra-cavity dipole trap, similar to the case depicted in Fig. 2.2a. Thus, the laser intensity strongly depends on the exact atomic position. Second, fluctuations of the Raman laser frequency with respect to that of the cavity will lead to a change in intensity. Both effects lead to a fluctuating Rabi frequency of the driven transition, which prevents controlled atomic state rotations when the Raman laser is impinging along the cavity axis. Therefore, to couple the levels $|1, 1\rangle$ and $|2, 2\rangle$, a different configuration has been used in this thesis, where both Raman laser beams impinge from the side of the cavity. One of the beams is polarized along the resonator axis, driving π transitions. The other beam is orthogonally polarized, thus equally driving σ^+ and σ^- transitions. While the σ^+ coupling to $|2, 2\rangle$ is desired, the σ^- component coupling to the state $|2, 0\rangle$ has to be eliminated. To this end, a magnetic field of about $0.5\,\text{G}$

is applied along the cavity axis. Due to the Zeeman effect, the atomic ground-state levels are thus shifted by $m_F \cdot 0.3$ MHz. In this way, the individual transitions can be spectrally resolved, which allows to drive only the desired transition, as described in more detail in Sect. 3.2.3.

3.2.3 Selective Addressing of Individual Zeeman States

In the following section, the use of a magnetic field to selectively drive transitions between individual atomic Zeeman levels is demonstrated. When a magnetic field of 0.5 G is applied along the quantization axis, the atomic Zeeman levels are shifted by $\pm 0.3\,\text{MHz} \cdot m_F$, as depicted in Fig. 3.4a. Here, the plus (minus) sign holds for the states in the $F = 2$ ($F = 1$) manifold. In order to drive the transition from $|1, 1\rangle$ to $|2, 2\rangle$ (yellow arrow), the polarization of one of the Raman beams is chosen to be linear and parallel to the cavity axis, corresponding to π transitions in the level scheme. The polarization of the other laser beam is set orthogonal to that of the first, thus driving $(\sigma^+ + \sigma^-)$ transitions. It follows that in this polarization configuration, $\Delta m_F = \pm 1$, such that six different Raman transitions can be addressed, which are shown as colored arrows in Fig. 3.4. In two cases, there are two transitions with degenerate frequency (blue, green). Thus, one expects to see only four distinct lines in a spectral measurement, where the atom is prepared in a random Zeeman state in the $F = 1$ manifold and the transfer probability upon irradiation of a Raman pulse is measured as a function of the detuning. The resulting graph can be seen in Fig. 3.4b.

One can clearly resolve four peaks, indicated by the arrows with the same color coding as in Fig. 3.4a. In spite of the fact that two degenerate transitions contribute to the center lines, the outer two peaks are a little higher, predominantly due to

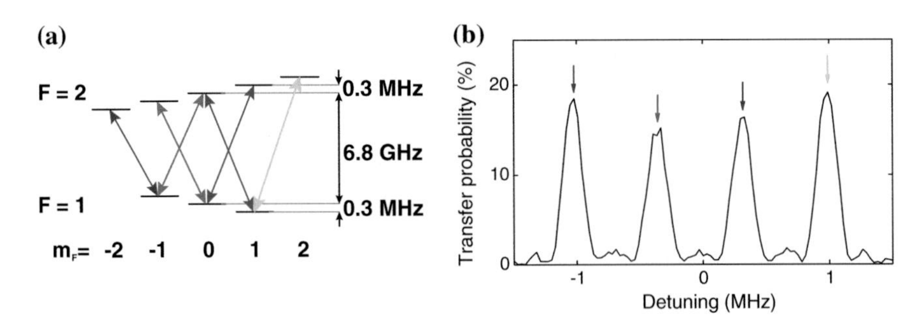

Fig. 3.4 **a** Shifts of the ^{87}Rb ground-state levels when a magnetic field of 0.5 G is applied along the quantization axis. With the used Raman laser configuration (orthogonal linear polarizations), only the transitions with $\Delta m_F = \pm 1$ (*colored arrows*) can be driven. **b** Raman spectroscopy of the atomic Zeeman sub-states. Due to the geometry and polarization of the Raman beams, one observes four separate lines, indicated by the arrows with the same color coding as in (**a**). The linewidth and the small side peaks are caused by the Fourier spectrum of the used square Raman pulse

much larger transition strengths. The smaller side-peaks observed in the spectrum are caused by the Fourier spectrum of the square Raman pulse of 5 μs duration. The spacing between adjacent transitions is 0.60 MHz. Therefore, the pulse duration of the Raman $\frac{\pi}{2}$ pulses employed in Chaps. 5 and 6 [2, 6] has been set to 1.7 μs, such that the first minimum of the Fourier spectrum coincides with the frequency of the adjacent transition. In this way, it is possible to selectively address the transition from $|1, 1\rangle$ to $|2, 2\rangle$ without driving that to $|2, 0\rangle$.

This property can also be applied to optically pump the atom to a desired state, by applying Raman lasers and a repumping laser to deplete all other states. As an example, consider optical pumping to the state $|1, 1\rangle$, which is again especially relevant for this thesis as it is part of the level scheme employed in Chaps. 5 and 6 [2, 6]. To deplete the states $|1, 0\rangle$ and $|1, -1\rangle$, a single Raman frequency is sufficient, which corresponds to the two green arrows in Fig. 3.4. Applying an additional laser resonant to the transition from $F = 2$ to the excited state $F' = 1$ leads to an accumulation of the atomic population in the state $|1, 1\rangle$. This can be seen when a spectrum is measured after pumping, see Fig. 3.5a (black line). The left two peaks are clearly suppressed compared to the right ones. For comparison, another spectrum was recorded when the atom had been pumped to $|1, -1\rangle$ using the same technique (blue line). Further optimization of the parameters has led to a pumping efficiency of about 90 %, which is unfortunately a rather low value, especially compared to the value of 97 % achieved when pumping to the state $|2, 2\rangle$ using a resonant laser along the cavity axis (as in Sect. 3.1). Nevertheless, the optical pumping scheme described above might be required in future experiments with several atoms in the same cavity mode: In this case, optical pumping along the cavity axis is expected to be hampered, as the pumping laser intensity in the cavity is strongly suppressed once one of the atoms is in the strongly coupled final state.

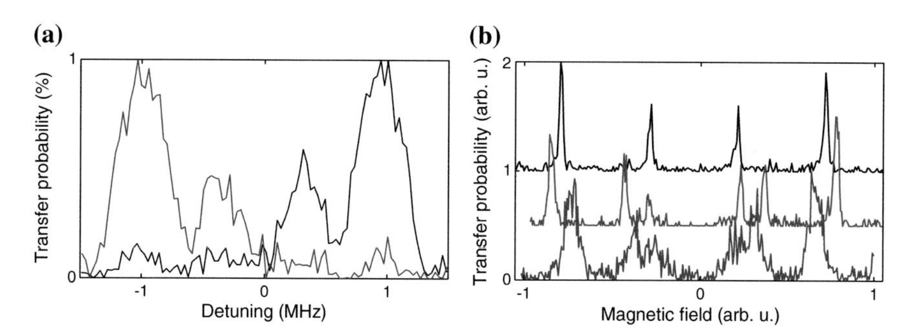

Fig. 3.5 **a** Optical pumping using the Raman lasers and an additional repumper. One can clearly observe a change in the atomic population, visible as the area under the *left* or *right* peak in the measured spectrum, when driving Raman transitions with a detuning of +300 kHz (*blue*) or −300 kHz (*black*). **b** Raman transfer of the atomic population when the detuning is set to a fixed frequency and the magnetic field is scanned along the cavity axis (*black*), the axis of the red-detuned dipole trap (*red*) or the blue-detuned vertical trap (*blue*)

The state-selective addressing of individual atomic transitions with Raman beams can also be used to obtain additional information about the physical system. In the following, the stability of the applied magnetic field is characterized by measuring the linewidth of the transitions between different hyperfine ground states. After preparing the atom in $F = 1$ a faint Raman pulse is irradiated for 300 μs, which ensures a narrow Fourier spectrum of the pulse. Then, the transfer probability to $F = 2$ is measured while the applied magnetic field is scanned along all three axes around its zero value. This gives several peaks in the spectrum, as can be seen in Fig. 3.5b. Along the cavity axis (black line), only four peaks are observed due to the polarization of the Raman beams (as explained in Sect. 3.2.1). Along the other axes, there are six peaks, which exhibit a much broader linewidth than the black curve.

When a magnetic field $B_{0,x}$ is applied along a certain axis x, line broadening $\Delta\nu$ due to fluctuations $\Delta B_{y,z} \ll B_{0,x}$ of the field along the perpendicular axes y, z is suppressed, as $\nu \propto |\mathbf{B}| = \sqrt{(B_{o,x} + \Delta B_x)^2 + \Delta B_y^2 + \Delta B_z^2} \simeq B_{o,x} + \Delta B_x$, and thus $\Delta\nu \propto \Delta B_x$. From the depicted data, one can thus conclude that the magnetic field fluctuations along the cavity axis are much smaller than those along the other axes, especially along the vertical axis (blue). A possible explanation is that the experimental setup is built on an optical table that consists of magnetizable steel. For operating the magneto-optical trap, a considerable magnetic quadrupole field is applied using a pair of horizontally oriented coils in anti-Helmholtz configuration. This might lead to shot-to-shot fluctuations of the magnetic field which are most prominent in the vertical direction.

Apart from magnetic fields, however, also circular components of the optical lattice light can lead to differential shifts of the atomic Zeeman levels via the AC Stark effect [16, 17] when the laser intensity fluctuates or the atom moves in the optical potential. Therefore, the polarization of the trap light is linearized with a measured extinction ratio of $10^6 : 1$. Nevertheless, the polarization of the trap might acquire circular components in the optical setup behind the polarizer. This effect is expected to be most severe along the cavity axis, as we realized during the characterization measurements of the quantum-memory experiment [18] that the cavity exhibits two orthogonal, linearly polarized eigenmodes, which are not fully degenerate in frequency but differ by about 400 kHz [19]. This can lead to circular polarization components within the cavity when the polarization axis of the trap does not coincide with one of the cavity eigenmodes. To minimize this effect, a magnetic field has been applied along the cavity axis to resolve the individual transitions. Subsequently, several Raman spectra have been recorded (as described above) with different orientation of the input polarization, see Fig. 3.6a. The minimally achieved linewidth (red data and red Lorentzian fit curve) on the $|1, 1\rangle \leftrightarrow |2, 2\rangle$ transition is 4 kHz. When the input polarization is rotated by five degrees to the left (black) or the right (blue) side, a shift and a clear broadening of the linewidth is observed.

In a second series of experiments, the same measurement has been performed with similar results for the polarization of the vertically oriented dipole trap (Fig. 3.6b). Here, the origin of the circular polarization components is different: For geometric reasons, the laser beam that forms the vertical trap first passes the polarizer and is

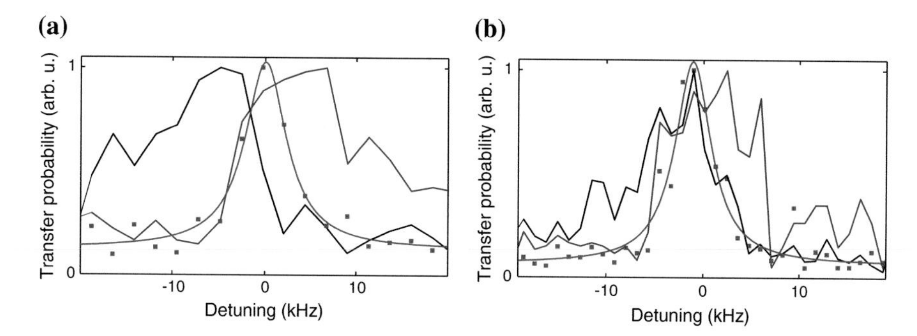

Fig. 3.6 Measurement of the Raman linewidth when the polarization of the trapping laser along the cavity axis (**a**) or along the vertical axis (**b**) is rotated by five degrees to one side (*black*) or the other (*blue*). Due to circular components of the trap light, the linewidth is broadened compared to the unrotated case (*red* data and Lorentzian fit curve)

then reflected upwards under an angle of 90°. If the polarization orientation is not exactly aligned to the plane in which the beam is reflected, then circular polarization components can be the result.

As a result from the measurements described above, the polarization of all trapping beams has been fixed at the optimum values. The observed linewidth allows for coherence times T_2^* on the order of 0.1 ms, which is sufficient for the experiments [2, 6] presented in Chaps. 5 and 6. Experiments that require longer coherence times will have to use dynamical decoupling [20–22] or the atomic clock states $|1,0\rangle$ and $|2,0\rangle$, because a state with $m_F = 0$ does not shift (to first order) in a small magnetic field. This property can be directly observed in the present setup, where a narrower linewidth of 1.5 kHz is observed on the atomic transition $|1,1\rangle \leftrightarrow |2,0\rangle$.

3.2.4 Atomic State Rotations

While the spectral addressing of the transition from $|\downarrow\rangle \equiv |1,1\rangle$ to $|\uparrow\rangle \equiv |2,2\rangle$ has been shown in Sect. 3.2.3, the ability to initialize the atomic state has been demonstrated in Sect. 3.1. Combining the two experimental techniques thus allows to fully control the atomic state, i.e. to prepare any superposition state $\alpha|\downarrow\rangle + \beta|\uparrow\rangle$, with α and β being normalized, complex parameters. In this context, the effect of the Raman laser beams is best visualized using the Bloch sphere to represent the atomic state in a two-level picture. Then, the atomic state $|\uparrow\rangle$ ($|\downarrow\rangle$) is located at the top (bottom) of a sphere, where the states $\frac{1}{\sqrt{2}}(|\downarrow\rangle + e^{i\phi}|\uparrow\rangle)$ lie on the equator, see Fig. 3.7a. The coordinate system is defined to rotate around the z axis with an angular frequency that equals $\Delta E/\hbar$, where $\Delta E = E_\uparrow - E_\downarrow$ is the energy difference between the two atomic levels. Then, the resonant Raman beams lead to a rotation of the atomic state on this sphere.

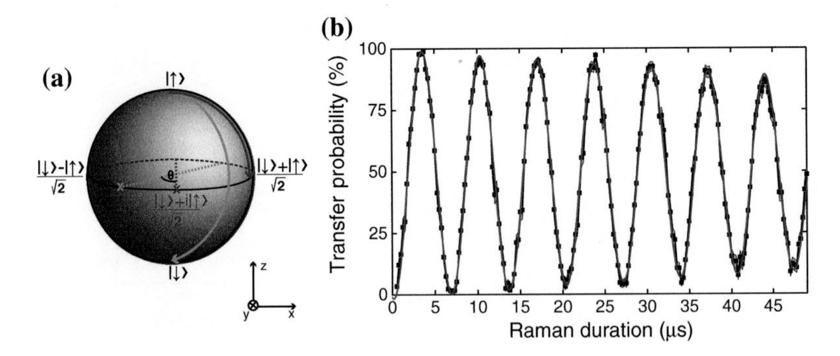

Fig. 3.7 **a** Visualization of the atomic state on the Bloch sphere. Any superposition state $\alpha\,|{\downarrow}\rangle + \beta\,|{\uparrow}\rangle$ is represented as a point on the surface of this *sphere*. The action of the Raman beams is to induce a rotation (*solid red* and *blue lines*) of the atomic state on the surface of the *sphere*. The rotation axis (*red* and *blue dashed lines*) is determined by the relative phase θ between the Raman lasers. **b** Rabi oscillations of the atomic population. The atom is prepared in the state $|{\uparrow}\rangle$ and the Raman lasers are applied for a varying duration. Then, the transfer probability to $|{\downarrow}\rangle$ is measured. The *red damped sinusoidal curve* is a theory fit to the data, which gives a damping time constant of 0.2 ms and a visibility of 98 %

The rotation axis lies in the equatorial plane and depends on the relative phase θ between the Raman lasers. The rotation speed is determined by the Rabi frequency Ω, which is proportional to the product of the intensities of the two lasers. The total rotation angle Φ can thus be controlled by adjusting the pulse area $\int \Omega(t)\mathrm{d}t$. As examples, two rotations $R_\theta(\Phi)$ around different axes are drawn: First, a rotation $R_0(\frac{\pi}{2})$ of $\Phi = 90°$ around the y axis ($\theta = 0$, red). Second, a rotation $R_\theta(\pi)$ of $\Phi = 180°$ around an axis determined by θ (light blue). In the experimental setup, θ can be directly set by changing the phase of a radio-frequency source (Analog Devices AD9910), which drives an AOM to generate one of the Raman laser beams.

To demonstrate full control over the atomic state, it is first prepared in $|{\uparrow}\rangle$. Subsequently, the Raman lasers are applied with a constant intensity for a varying duration. Then, the probability to transfer the state to $|{\downarrow}\rangle$ is measured and the result is shown in Fig. 3.7b. One can clearly observe a sinusoidal oscillation from 0 % close to the expected 100 %. From the red fit curve, a visibility of 98 % is inferred, which is limited by the quality of the state initialization and readout (compare Sect. 3.1). The observed decay of the oscillation with a time constant of 0.2 ms is due to decoherence, which is further investigated in the following using Ramsey spectroscopy.

To this end, the atom is prepared in $|{\uparrow}\rangle$ and a rotation $R_0(\pi/2)$ is applied, transferring the atomic state to $(|{\downarrow}\rangle + |{\uparrow}\rangle)/\sqrt{2}$. In the rotating frame, this state remains static on the Bloch sphere when the Raman laser are subsequently switched off. After a certain dark time interval τ, the lasers are switched back on, again leading to a rotation $R_\theta(\pi/2)$. When the phase θ between the laser beams has the same value as before, the rotation is around the same axis, such that the atom ends up in the state $|{\downarrow}\rangle$. If, however, $\theta = \pi$, the atom will be rotated back to the state $|{\uparrow}\rangle$, as $R_\pi(\pi/2) = R_0(-\pi/2)$.

The reason for a phase change $\Delta\theta$ can be that the Raman lasers operate at a small detuning δ with respect to the frequency difference of the atomic states $\Delta E/\hbar$. In this case, the phase acquired in τ will be $\delta \cdot \tau$. This effect can be seen in Fig. 3.8, where the detuning δ has been scanned for two different values of $\tau = 1\,\mu s$ (a) and $\tau = 6\,\mu s$ (b). One can clearly observe a sinusoidal modulation of the transfer probability, multiplied with a $(\sin(x)/x)^2$ envelope that stems from the Fourier spectrum of the applied square Raman $\pi/2$ pulses. As expected, the modulation has a smaller period when the hold time is increased. Thus, further increasing τ allows one to determine the atomic transition energy difference with a very high accuracy.

This is demonstrated in Fig. 3.8c, where the hold time has been increased to 20 μs (red) or 30 μs (blue). From the sinusoidal fit curves, the atomic transition frequency is determined; it is detuned from the ground-state hyperfine splitting by 1.07 MHz. The achievable resolution, however, is limited by decoherence. More specifically, fluctuations of the atomic transition frequency $\Delta\omega(t)$, e.g. due to a fluctuating magnetic field, can lead to a random phase of $\int_0^\tau \Delta\omega(t) \cdot t\,\mathrm{d}t$ upon application of the second Raman pulse. This results in an exponential decay $\propto \exp(-\frac{\tau}{T_2^*})$ of the fringe visibility with increasing hold time. Thus, the Ramsey technique can be employed to directly measure this dephasing time T_2^*. We again find a value on the order of 0.1 ms, in good agreement with the results of the linewidth measurements presented in Sect. 3.2.3. The high degree of control over the atomic state, however, allows to partially undo the dephasing by a technique called spin echo, which has been pioneered in nuclear magnetic resonance (NMR) experiments and has meanwhile found numerous applications in quantum control [23].

The working principle of this technique is to apply an additional Raman rotation $R_{\pi/2}(\pi)$ at $t = \tau/2$. The effect of this pulse is a transfer of the state $(|\downarrow\rangle + \mathrm{e}^{i\phi(t)}\,|\uparrow\rangle)/\sqrt{2}$ to the state $(|\downarrow\rangle - \mathrm{e}^{-i\phi(t)}\,|\uparrow\rangle)/\sqrt{2}$, which includes a sign change of the acquired phase difference $\phi(\tau/2) = \int_0^{\tau/2} \Delta\omega(t) \cdot t\,\mathrm{d}t$. Thus, after the full hold time $\phi(\tau) = -\int_0^{\tau/2} \Delta\omega(t) \cdot t\,\mathrm{d}t + \int_{\tau/2}^{\tau} \Delta\omega(t) \cdot t\,\mathrm{d}t$. If the fluctuations of the atomic transition frequency are on a timescale that is large compared to the hold time, i.e. $\Delta\omega(t) \equiv \Delta\omega$, the two terms cancel and the full fringe visibility is recovered. Figure 3.8d shows the result of a respective measurement with $\tau = 40\,\mu s$. During the experiment, a static magnetic field B_0 of varying strength is applied. Without the spin echo pulse (blue), the atomic state acquires a phase $\propto \Delta B_0 \cdot \tau$, leading to the observed fringes. Again, the visibility is reduced due to dephasing.

When the spin echo pulse is applied (red), the dephasing is perfectly reversed for small values of ΔB_0, and the second Raman pulse transfers the atom back to its original state with a probability close to 100 %. Thus, the coherence time of the system can be strongly increased. Experimentally, we find a value of about 1 ms (data not shown), which is likely limited by dephasing effects on shorter time scales, which can not be perfectly reversed with the simple employed sequence. However, further improvement is likely possible with sequences that use several echo pulses [23], which can be optimized to achieve excellent dynamical decoupling [20–22, 24]. Ultimately, the coherence time will only be limited by other effects (such as trap-light scattering) that cannot be reversed at all.

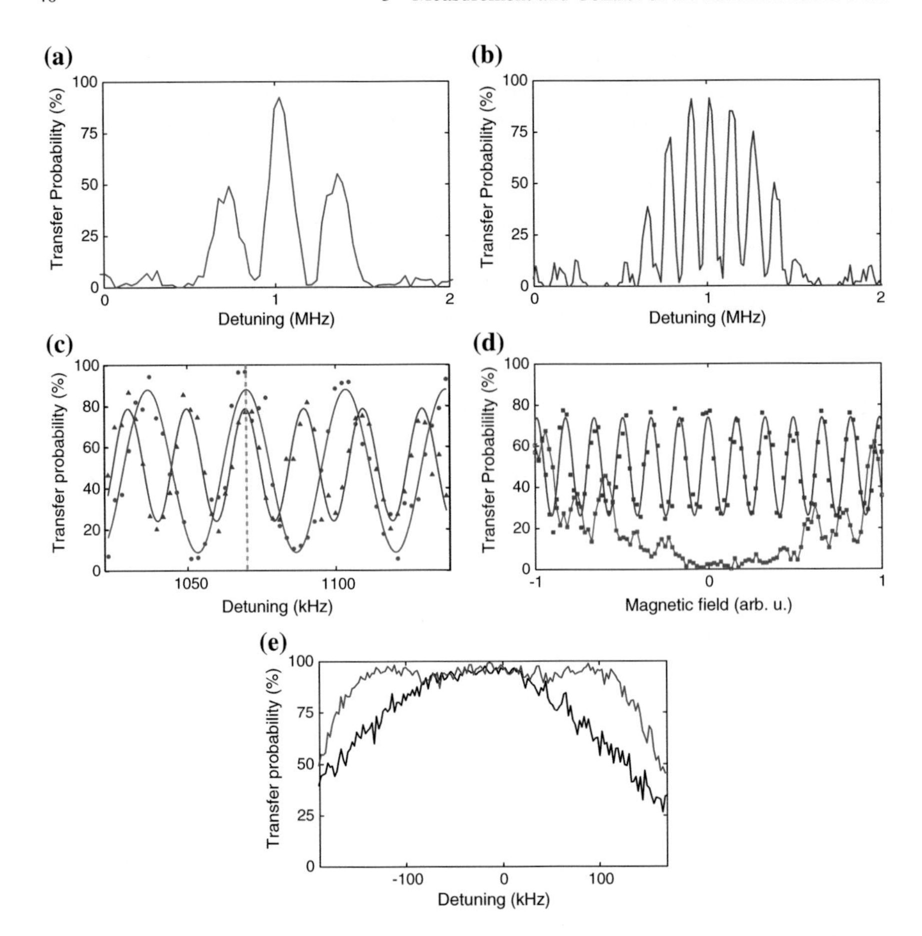

Fig. 3.8 **a–d** Ramsey spectroscopy. The atom is prepared in the state $|2, 2\rangle$ and transitions to the state $|1, 1\rangle$ are driven. These transitions are centered around a detuning of $\delta = 1$ MHz from the ground-state hyperfine splitting due to the Zeeman shift of the used atomic levels. In the depicted spectra, two $R_0(\pi/2)$ Raman pulses are applied with a delay of 1 μs (**a**) or 6 μs (**b**). A sinusoidal modulation is observed, whose period depends on the pulse delay, multiplied with a $\left(\frac{\sin(\delta)}{\delta}\right)^2$ envelope which stems from the Fourier spectrum of the applied Raman pulses. (**c**) Ramsey spectroscopy with longer delay times of 20 μs (*red*) or 30 μs (*blue*). The atomic transition frequency is at the detuning where all three *curves* exhibit a maximum in the transfer probability, i.e. at 1.07 MHz (*green dashed line*). The reduction in the amplitude of the oscillation is due to decoherence. (**d**) Spin echo measurement. The atom is prepared in the state $|2, 2\rangle$ and two $R_0(\pi/2)$ Raman pulses are applied with a delay of 40 μs. When the strength of the magnetic guiding field is varied along its direction, the phase of the atomic state is changed upon irradiation of the second Raman pulse. This leads to a oscillation of the transfer probability (*blue data and sinusoidal fit curve*), whose visibility is reduced due to Random phase fluctuations. This effect can be compensated (for small values of the magnetic field change) when a $R_0(\pi)$ Raman pulse is applied after half of the delay time (*red*). (**e**) Measurement of the transfer probability when using composite pulses. The atom is prepared in the state $|\uparrow\rangle$. As a reference, a $R_0(\pi)$ Raman rotation is applied (*black*) and the transfer probability to $|\downarrow\rangle$ is measured. When a composite pulse sequence, consisting of the rotations $R_0\,(\pi/2)\,R_{\pi/2}\,(\pi)\,R_0\,(\pi/2)$ (*red*), is employed, the transfer probability stays high over a larger range of detunings, making it more robust with respect to experimental imperfections

The spin echo employed above is not the only NMR technique that might prove useful in the context of atomic-state control. Also the use of composite pulses to improve the fidelity of the atomic state rotations can be directly implemented in our setup. The basic idea is that the simple Raman pulses employed above can be prone to several sources of errors. Prime examples are a residual detuning of the Raman lasers with respect to the atomic transition, or a jitter in the rotation angle caused by a fluctuation of the pulse area. These problems can be overcome to a large extent by using a specific series of laser pulses to implement the desired rotation. A well-known example is the operation $R_0(\pi/2) \cdot R_{\pi/2}(\pi) \cdot R_0(\pi/2)$, which implements a rotation $R_0(\pi)$ where the error of the individual pulses cancels to a large extent. Figure 3.8e shows the implementation of this pulse sequence. In this measurement, the atom is prepared in $|\uparrow\rangle$ before the rotation is applied. The detuning of the Raman beams is varied to mimic a drift in the atomic transition frequency. With a simple $R_0(\pi)$ rotation (black), the fidelity of the state transfer drops quadratically with the detuning. With the composite pulse (red), the range in which a high fidelity is obtained is considerably larger. In a similar way, the influence of drifts in the laser power is reduced.

While in principle, similar pulse sequences exist to implement $R_0(\pi/2)$ rotations, single pulses have been used in the experiments presented in Chap. 5 [2] and Chap. 6 [6], because it turned out in the measurement above that the fidelity of a single pulse is not improved considerably. Thus, the overhead of implementing such a composite sequence is not required in all experiments that involve only a few atomic state rotations.

References

1. C.-Y. Shih, M.S. Chapman, Nondestructive light-shift measurements of single atoms in optical dipole traps. Phys. Rev. A **87**(6), 063408 (2013), 00003. doi:10.1103/PhysRevA.87.063408. http://link.aps.org/doi/10.1103/PhysRevA.87.063408
2. A. Reiserer, S. Ritter, G. Rempe, Nondestructive detection of an optical photon. Science **342**(6164), 1349–1351 (2013). doi:10.1126/science.1246164. http://www.sciencemag.org/content/342/6164/1349
3. A. D. Boozer et al., Cooling to the ground state of axial motion for one atom strongly coupled to an optical cavity. Phys. Rev. Lett. **97**(8), 083602 (2006). doi:10.1103/PhysRevLett.97.083602. http://link.aps.org/doi/10.1103/PhysRevLett.97.083602
4. J. Bochmann et al., Lossless state detection of single neutral atoms. Phys. Rev. Lett. **104**(20), 203601 (2010), 00051. doi:10.1103/PhysRevLett.104.203601. http://link.aps.org/doi/10.1103/PhysRevLett.104.203601
5. R. Gehr et al., Cavity-based single atom preparation and high-fidelity hyperfine state readout. Phys. Rev. Lett. **104**(20), 203602 (2010). doi:10.1103/PhysRevLett.104.203602. http://link.aps.org/doi/10.1103/PhysRevLett.104.203602
6. A. Reiserer et al., A quantum gate between a flying optical photon and a single trapped atom. Nature **508**(7495), 237–240 (2014), 00010. ISSN: 0028-0836.doi:10.1038/nature13177. http://www.nature.com/nature/journal/v508/n7495/full/nature13177.html
7. W. Nagourney, J. Sandberg, H. Dehmelt, Shelved optical electron amplifier: Observation of quantum jumps. Phys. Rev. Lett. **56**(26), 2797–2799 (1986). doi:10.1103/PhysRevLett.56.2797. http://link.aps.org/doi/10.1103/PhysRevLett.56.2797

8. J.C. Bergquist et al., Observation of quantum jumps in a single atom. Phys. Rev. Lett. **57**(14), 1699–1702 (1986). doi:10.1103/PhysRevLett.57.1699. http://link.aps.org/doi/10.1103/PhysRevLett.57.1699
9. D. Leibfried et al., Quantum dynamics of single trapped ions. Rev. Mod. Phys. **75**(1), 281 (2003). doi:10.1103/RevModPhys.75.281. http://link.aps.org/doi/10.1103/RevModPhys.75.281
10. D.J. Heinzen, D.J. Wineland, Quantum-limited cooling and detection of radio-frequency oscillations by laser-cooled ions. Phys. Rev. A **42**(5), 2977–2994 (1990), 00230. doi:10.1103/PhysRevA.42.2977. http://link.aps.org/doi/10.1103/PhysRevA.42.2977
11. C. Monroe et al., Demonstration of a fundamental quantum logic gate. Phys. Rev. Lett. **75**(25), 4714–4717 (1995), 01678. doi:10.1103/PhysRevLett.75.4714. http://link.aps.org/doi/10.1103/PhysRevLett.75.4714
12. D.D. Yavuz et al., Fast ground state manipulation of neutral atoms in microscopic optical traps. Phys. Rev. Lett. **96**(6), 063001 (2006). doi:10.1103/PhysRevLett.96.063001. http://link.aps.org/doi/10.1103/PhysRevLett.96.063001
13. M.P.A. Jones et al. Fast quantum state control of a single trapped neutral atom. Phys. Rev. A **75**(4), 040301 (2007). doi:10.1103/PhysRevA.75.040301. http://link.aps.org/doi/10.1103/PhysRevA.75.040301
14. I.Dotsenko et al., Application of electro-optically generated light fields for Raman spectroscopy of trapped cesium atoms. Appl. Phys. B **78**(6), 711–717 (2004). ISSN: 0946–2171, 1432–0649.doi:10.1007/s00340-004-1467-9. http://link.springer.com/article/10.1007/s00340-004-1467-9
15. A. Reiserer et al., Ground-state cooling of a single atom at the center of an optical cavity. Phys. Rev. Lett. **110**(22), 223003 (2013). doi:10.1103/PhysRevLett.110.223003. http://link.aps.org/doi/10.1103/PhysRevLett.110.223003
16. R. Grimm, M. Weidemüller, Y. B. Ovchinnikov, Optical dipole traps for neutral atoms. Adv. At. Mol. Opt. Phys. **42**, 95-170 Academic Press (2000) ISBN: 978-0-12-003842-8. http://www.sciencedirect.com/science/article/pii/S1049250X0860186X
17. W. Rosenfeld et al, Coherence of a qubit stored in Zeeman levels of a single optically trapped atom. Phys. Rev. A **84**(2), 022343 (2011). doi:10.1103/PhysRevA.84.022343. http://link.aps.org/doi/10.1103/PhysRevA.84.022343
18. H.P. Specht et al., A single-atom quantum memory'. Nature **473**(7346), 190–193 (2011), 00128. ISSN: 0028-0836.doi:10.1038/nature09997. http://dx.doi.org/10.1038/nature09997
19. H. Specht, Einzelatom-quantenspeicher für polarisations-qubits. PhD thesis. Technische Universität München, 2010. http://mediatum.ub.tum.de/node?id=1002627
20. L. Viola, E. Knill, S. Lloyd, Dynamical decoupling of open quantum systems. Phys. Rev. Lett. **82**(12), 2417–2421 (1999). doi:10.1103/PhysRevLett.82.2417. http://link.aps.org/doi/10.1103/PhysRevLett.82.2417
21. G.S. Uhrig, Keeping a quantum bit alive by optimized pi-pulse sequences. Phys. Rev. Lett. **98**(10), 100504 (2007). doi:10.1103/PhysRevLett.98.100504. http://link.aps.org/doi/10.1103/PhysRevLett.98.100504
22. M.J. Biercuk et al., Optimized dynamical decoupling in a model quantum memory. Nature **458**(7241), 996-1000 (2009). ISSN: 0028-0836. doi:10.1038/nature07951. http://www.nature.com/nature/journal/v458/n7241/full/nature07951.html
23. L.M.K. Vandersypen, I.L. Chuang, NMR techniques for quantum control and computation. Rev. Mod. Phys. **76**(4), 1037–1069 (2005). doi:10.1103/RevModPhys.76.1037. http://link.aps.org/doi/10.1103/RevModPhys.76.1037
24. G.de Lange et al., Universal dynamical decoupling of a single solid-state spin from a spin bath. Science **330**(6000), 60–63 (2010). ISSN: 0036-8075, 1095-9203. doi:10.1126/science.1192739. http://www.sciencemag.org/content/330/6000/60

Chapter 4
Controlled Phase Gate Mechanism

4.1 Working Principle

In this section, the interaction mechanism presented in this thesis to realize a controlled phase gate is analyzed in more detail. In particular, the reflection process that causes the state-dependent phase shift of π is described mathematically. In contrast to the simplified, but intuitive model introduced in Sect. 1.2, the effect of additional cavity losses and finite coupling strength will be considered in the following. To this end, the common description of a strongly-coupled atom-cavity system that is based on the Jaynes-Cummings model [1] and includes the coupling to the outside world as a damping term [2], has to be extended using the framework of input-output-theory [3–5]. Consider the case of an asymmetric cavity with field decay rates κ_r and κ_l through the right and left mirror and losses κ_{loss} (see Fig. 4.1). The operators $a_{\text{in,l}}$ ($a_{\text{in,r}}$) are the field operators of an electromagnetic mode impinging from the left (right) side, while $a_{\text{out,l}}$ ($a_{\text{out,r}}$) describe the outgoing field to the left (right) side, respectively. a is the annihilation operator for a photon in the cavity field. Then the relation between the input and output modes (in the Heisenberg picture) is given by [3–5]:

$$a_{\text{out,r}}(t) + a_{\text{in,r}}(t) = \sqrt{2\kappa_r}\, a(t) \tag{4.1}$$

$$a_{\text{out,l}}(t) + a_{\text{in,l}}(t) = \sqrt{2\kappa_l}\, a(t) \tag{4.2}$$

In the following, it is assumed that an atom is trapped in the cavity, on resonance with a strongly coupled atomic dipole transition between a ground state $|\text{g}\rangle$ and an excited state $|\text{e}\rangle$. The Hamiltonian H which describes the coherent interaction between the atom and the cavity is then:

$$H = \hbar g \left(|\text{e}\rangle \langle \text{g}|\, a + |\text{g}\rangle \langle \text{e}|\, a^\dagger \right) \tag{4.3}$$

A. Reiserer, *A Controlled Phase Gate Between a Single Atom and an Optical Photon*, Springer Theses,
DOI 10.1007/978-3-319-26548-3_4

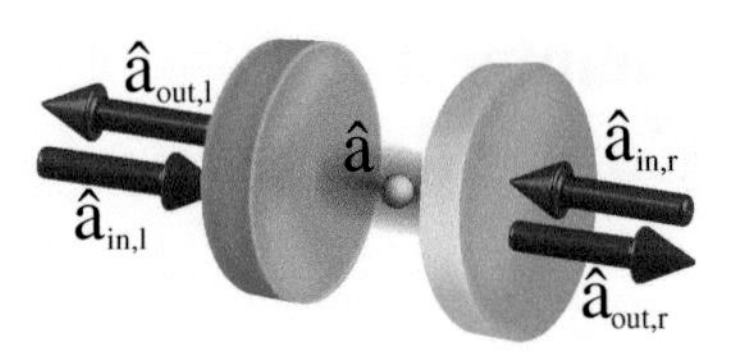

Fig. 4.1 Schematic representation of the input and output operators of the cavity field

Assume the system is weakly driven from the left side only. Then the above relations allow to calculate the reflection and transmission properties by solving the Heisenberg-Langevin equation of motion [4–6]:

$$\dot{a} = -\frac{i}{\hbar}[a, H] - \kappa a + \sqrt{2\kappa_l} a_{\text{in,l}}(t) \tag{4.4}$$

In this equation, $\kappa = \kappa_l + \kappa_r + \kappa_{\text{loss}}$ is the total cavity field decay rate. Equation (4.4) can be integrated numerically to calculate the dynamics of the system. However, an analytic result can also be obtained [7, 8] when the atomic excitation is negligibly small and the photonic wavepacket envelope only varies on a long timescale. This results in the following relations for the amplitude reflection $r(\omega)$ and transmission $t(\omega)$ coefficients, depending on the detuning $\Delta_{c(a)} \equiv \omega - \omega_{c(a)}$ of the driving laser with respect to the cavity (atomic transition) frequency ω_c (ω_a), respectively:

$$r(\omega) = 1 - \frac{2\kappa_l(i\Delta_a + \gamma)}{(i\Delta_c + \kappa)(i\Delta_a + \gamma) + g^2} \tag{4.5}$$

$$t(\omega) = \frac{2\sqrt{\kappa_l \kappa_r}(i\Delta_a + \gamma)}{(i\Delta_c + \kappa)(i\Delta_a + \gamma) + g^2} \tag{4.6}$$

The phase of the reflected light is $\arg(r(\omega))$, while the intensity reflection and transmission coefficients are given by $R(\omega) = |r(\omega)|^2$ and $T(\omega) = |t(\omega)|^2$, respectively. For the empty cavity ($g = 0$) these relations become:

$$R(\Delta_c) = 1 - \frac{4\kappa_l(\kappa_r + \kappa_{\text{loss}})}{\Delta_c^2 + \kappa^2} \tag{4.7}$$

$$T(\Delta_c) = \frac{4\kappa_l \kappa_r}{\Delta_c^2 + \kappa^2} \tag{4.8}$$

Thus, both the transmission and reflection spectra exhibit a Lorentzian shape of FWHM 2κ. On resonance, the transmission is $T(\Delta_c = 0) = \frac{4\kappa_l\kappa_r}{\kappa^2}$, and the reflection is $R(\Delta_c = 0) = 1 - \frac{4\kappa_l(\kappa_r+\kappa_{\text{loss}})}{\kappa^2}$. Measuring the spectrum of the cavity transmission and reflection allows to quantify the rates κ_l, κ_r and κ_{loss} and thus to measure the transmission of the cavity mirrors and the combined absorption and scattering losses. However, the influence of the mode-matching efficiency ξ between the transversal profile of the impinging light and that of the cavity mode has so far been neglected. In our experiment, ξ can be estimated by measuring the coupling efficiency of the light transmitted through the cavity into an anti-reflection coated optical fiber. Typical values of $\xi = 92(2)\,\%$ are achieved, making the above equations still a good approximation. A better estimation, however, is possible by setting $R_{\mathrm{exp}}(\Delta_c) = (1-\xi) + \xi \cdot R(\Delta_c)$, as any light field that is not coupled to the cavity will be reflected with $R \simeq 1$.

Figure 4.2 shows reflection (a) and transmission (b) spectra, measured in our experimental setup in the coupled (red) and uncoupled (blue) case. The solid blue lines are fit curves according to the equations above, which yield $\kappa = 2\pi \cdot 2.5(1)$ MHz and $\kappa_l = 2\pi \cdot 2.3(1)$ MHz. With the mirror separation of 0.486 mm, the individual losses can be calculated. The round-trip loss is 102(3) ppm, in excellent agreement with an earlier ring-down measurement [9] that gave 103(2) ppm. The transmission of the outcoupling mirror is $T_l = 92(3)$ ppm, slightly different from what has been measured before the cavity was assembled (99 ppm) [9]. From the ratio of absorption and transmission, it furthermore follows that $T_r = 3(1)$ ppm, Loss $= 7(3)$ ppm.

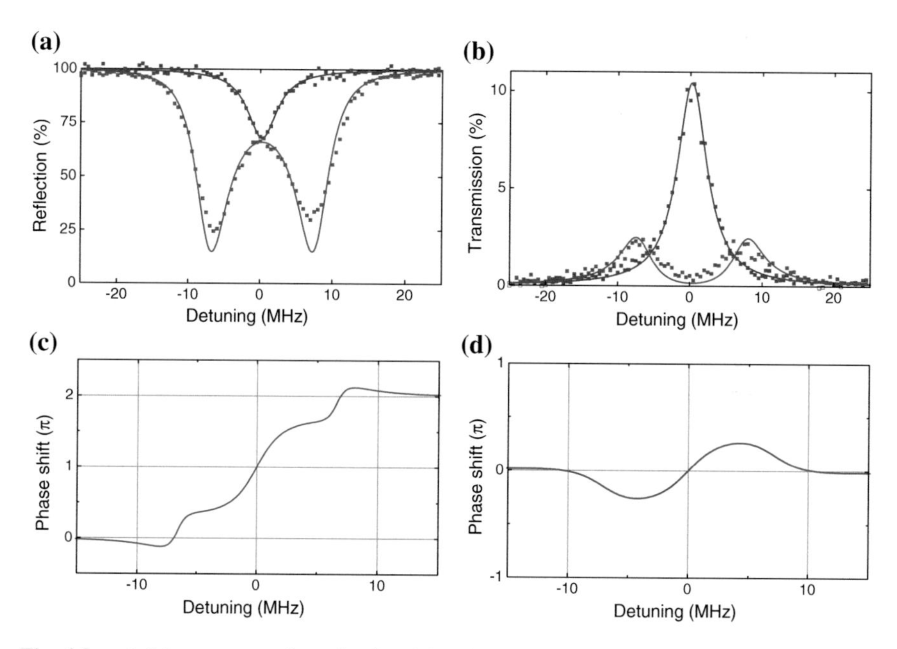

Fig. 4.2 **a, b** Measurement of a reflection (**a**) and transmission (**b**) spectrum, with the atom in a coupled (*red*) or uncoupled (*blue*) state. The blue Lorentzian fit curves are used to determine the cavity parameters. With this, the red theory curves (see text) are calculated, which show good qualitative agreement with the measured data. **c, d** Phase difference (*red*) of the reflected (**c**) and transmitted (**d**) light field between the coupled and uncoupled case, calculated for the measured parameters of the experimental setup. In a small frequency window around zero detuning, a relative phase shift of π is expected in reflection

The red curves are calculated using $\xi = 0.9$, $\Delta_a = \Delta_c$, the fit values of κ and a coupling strength $g = 2\pi \cdot 7\,\text{MHz}$. Good qualitative agreement with the measured data is observed. The deviation, however, is quite considerable on the normal modes in the reflection spectrum. This can be explained by a fluctuating AC Stark shift ($\Delta_a \neq \Delta_c$) of a few MHz, compatible with independent measurements (see Sect. 2.4), that broadens and flattens the observed dips.

With respect to the atom-photon interaction mechanism, the most important result from this measurement is the reflection on resonance (i.e. at $\Delta_c = 0$). In 30 % of the cases, an impinging photon is lost. It is either scattered by the atom or transmitted or absorbed within the cavity. In the remaining 70 %, the interaction mechanism is expected to work as intended, as can be seen from a calculation of the expected phase response of the system. Using the results from above gives the red curve in Fig. 4.2c (reflection) and d (transmission). The transmission curve is in good agreement with the results of a recent publication [10]. In reflection, a phase shift of π is expected around zero detuning, meaning that the interaction mechanism works as intended. However, it is important to operate in a small frequency window around this value. This is investigated in more detail in the following Sect. 4.2.

4.2 Bandwidth

This section will provide additional insight into the frequency dependence of the used atom-photon interaction mechanism. Note that the steady-state solution presented in Sect. 4.1 assumes impinging photons with a very narrow spectral distribution $\Delta\nu$, i.e. a very long temporal duration Δt. However, from an experimental point of view one is interested in using photons that are as short as possible, mainly to minimize decoherence during the reflection process. Therefore, the bandwidth of the used atom-photon interaction mechanism is investigated in the following.

To this end, a first measurement is performed, where faint laser pulses are reflected from the cavity. As explained in detail in Chap. 6, in the employed level scheme (see Fig. 6.1) only right-circularly polarized photons ($|R\rangle$) are strongly coupled to the atom in the state $|2, 2\rangle$, while the coupling is negligible for left-circular polarization ($|L\rangle$) due to a relative detuning between the excited atomic states $\left|3', 3\right\rangle$ and $\left|3', 1\right\rangle$. This ideally leads to a flip of any linear input polarization $(|R\rangle + e^{i\phi}\,|L\rangle)/\sqrt{2} \rightarrow (|R\rangle - e^{i\phi}\,|L\rangle)/\sqrt{2}$ upon reflection from the setup. In Fig. 4.3, the measured probability to observe this polarization flip is shown upon stepwise reduction of the photon duration. Above a duration of about 0.5 μs, this probability remains constant, which means that the atom-photon interaction mechanism works as intended. Below this value, the polarization flip probability breaks down quite rapidly.

Additional information about the bandwidth can be obtained in spectral measurements. To this end, the atom is prepared in the coupled state and photons in the input state $|D\rangle$ are reflected from the setup, which have a duration of several μs and correspondingly a narrow spectral distribution (spectral width $\Delta\nu \simeq 0.1\,\text{MHz}$). The

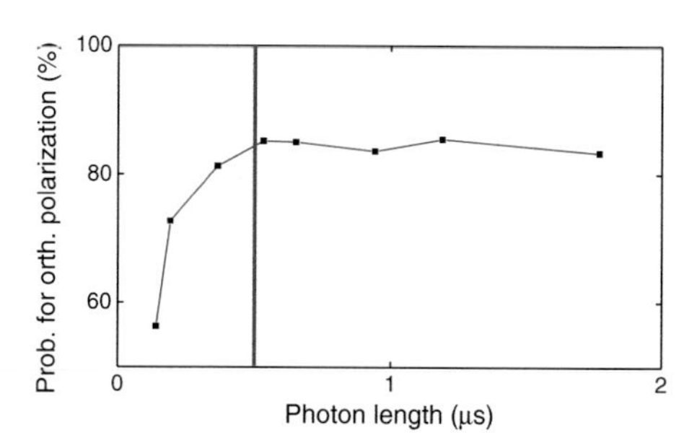

Fig. 4.3 Bandwidth of the atom-photon interaction mechanism. Weak coherent laser pulses of linear polarization are reflected and measured in a polarization-resolving setup. The *black* data show the probability of observing a polarization flip, depending on the duration of the impinging photons (FWHM of the measured Gaussian temporal envelope). Below 0.5 μs (*blue line*), a steep decrease is observed

central frequency of these pulses is scanned while the cavity and atomic frequencies remain fixed. Measuring the polarization flip probability as defined above gives the black curve in Fig. 4.4a. The observed value remains constant in a range of about 2 MHz, consistent with the results of the temporal measurement (Fig. 4.3).

The polarization state of the reflected light can be fully reconstructed by measuring in three orthogonal polarization bases, say horizontal/vertical ($|H\rangle = (|R\rangle + |L\rangle)/\sqrt{2}$ and $|V\rangle = (|R\rangle - |L\rangle)/\sqrt{2}$), diagonal/antidiagonal ($|D\rangle = (|R\rangle + i\,|L\rangle)/\sqrt{2}$ and $|A\rangle = (|R\rangle - i\,|L\rangle)/\sqrt{2}$), and right/left-circular. While the black curve in Fig. 4.4a shows the ratio of $p_A = \frac{P_{|A\rangle}}{P_{|D\rangle}+P_{|A\rangle}}$, the blue curve shows $p_V = \frac{P_{|V\rangle}}{P_{|H\rangle}+P_{|V\rangle}}$ and the red curve $p_L = \frac{P_{|L\rangle}}{P_{|R\rangle}+P_{|L\rangle}}$, all for the same input state $|D\rangle$.

The red curve that corresponds to the circular polarization components stays constant over a comparably large frequency range, which is expected as the setup has almost equal losses in the coupled and uncoupled case. Thus, the polarization of any resonant, linearly polarized input photon also stays linear upon reflection from the system. The increase at larger detunings is due to absorption of the right-circular polarization components on the normal mode (see also Fig. 4.2a).

The measurement in all three polarization bases allows to fully reconstruct the photonic state and thus to determine the phase shift φ between the right-circular polarization component and the left-circular one. In the range where the red curve in Fig. 4.4a is flat and the photonic state remains close to the equator of the Bloch sphere, φ can be calculated using $\tan\varphi = \frac{2p_H-1}{2p_D-1}$. Figure 4.4b shows the data obtained in this way; one observes good agreement with a theory curve (red) that has been calculated (as explained in Sect. 4.1) using the measured system parameters, shifted by a constant offset of 0.6 MHz. This offset is partially due to cavity birefringence: The reflected state was $|D\rangle$, while the cavity was stabilized on resonance with $|A\rangle$. The measured shift is thus comparable to an independent characterization of the relative frequency shift of the polarization eigenmodes of the cavity, 0.4 MHz [11].

The atom-photon interaction mechanism presented in this work relies on a constant phase shift of π between the coupled and the uncoupled component, which is observed

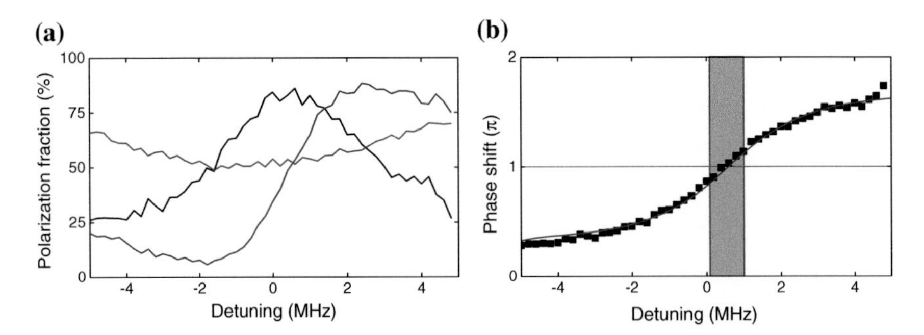

Fig. 4.4 **a** Rotation of the photonic polarization when the atom is prepared in the coupled state and a D-polarized photon is reflected from the system and subsequently measured in a polarization-resolving setup in the A/D basis (*black*), the R/L basis (*red*) and the H/V basis (*blue*). The phase shift and thus the polarization rotation strongly depends on the photon detuning. **b** Phase shift of the right-circular polarization component relative to the left-circular one. The *black* data points are obtained from the measurement. The *red line* is a theory curve, using the measured system parameters (see Sect. 4.1) and a frequency offset of 0.6 MHz. The *blue area* (width: 0.9 MHz) indicates the bandwidth of a Gaussian photon wavepacket (FWHM), at which a decrease in the polarization flip probability sets in, corresponding to the blue line in Fig. 4.3

in a narrow frequency window around the measured offset value of 0.6 MHz. The maximum width of this frequency window is determined by the cavity linewidth: $\Delta\nu \ll \kappa$. To understand this intuitively, consider the curve in Fig. 4.4b. The maximum bandwidth before the gate fidelity is severely reduced is shown as a blue rectangle (derived from the measurements of Fig. 4.3). In this regime, the phase dispersion is well approximated by a linear curve, with the slope determined by κ. A linear phase increase in the frequency domain is equivalent to a constant shift of the reflected photonic wavepacket in the time domain. The mechanism starts to fail when this shift becomes comparable to the pulse duration, i.e. when the wavepacket in the uncoupled case has a reduced temporal overlap with the coupled one.

The measurement of Fig. 4.4b also demonstrates the critical role of a possible frequency offset between the impinging photons and the cavity. When the detuning drifts over time or fluctuates on a short timescale, the phase shift will vary and thus the fidelity of the atom-photon interaction mechanism will be strongly affected. Therefore, the stability of this detuning is investigated in the following. In the experimental setup, the cavity is stabilized to the frequency of the intra-cavity dipole trap laser using the Pound-Drever-Hall technique [12]. This laser is in turn stabilized to an optical frequency comb. The faint pulses at 780 nm that are used to characterize the setup in reflection are also referenced to this comb. To determine the stability of the relative detuning, a strong continuous beam, derived from the 780 nm laser, is impinging onto the cavity. The beam has a detuning of 2.5 MHz with respect to the cavity frequency, such that it is located at a transmission of 50 % of the maximum value, i.e. on the side of the Lorentzian curve. The transmission through the cavity is measured with a photodiode. The standard deviation of this signal allows to deduce the stability σ_ν of the cavity frequency. After optimization of the locking electronics,

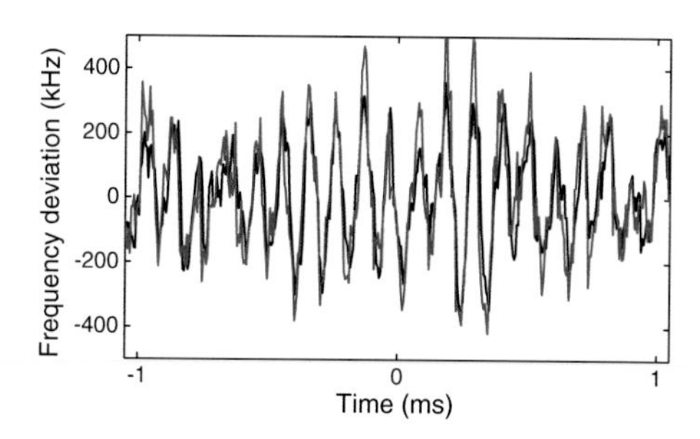

Fig. 4.5 Stability of the cavity frequency. The *black* signal is the error signal of the cavity lock, the *red* signal is the transmission of the probe laser (side of fringe). The *two curves* show a clear correlation

the resulting value has been improved in the course of this thesis by a factor of three to $2\sigma_\nu \simeq 0.3\,\text{MHz}$. This value is limited by the mechanical stability of the setup, as can be seen in Fig. 4.5, where the temporal fluctuations of the transmission (red) have been recorded with the optimized settings of the cavity lock. The measured amplitude fluctuations have been converted to a frequency deviation scale using the independently measured cavity linewidth and assuming a linear scaling, which is a good approximation at small excursions. The transmission oscillates at a frequency of about 10 kHz, which coincides with the first mechanical resonances of the piezo tube that holds the cavity mirrors [13]. In spite of the fact that the oscillation is also observed in the error signal of the locking electronics (black line), it could not be compensated by the lock due to the large phase shift of the mechanical response upon passing over a resonance. Thus, the achieved frequency stability on short timescales slightly reduces the fidelity of the atom-photon interaction mechanism, as the curves in Fig. 4.4b change considerably over a frequency range of 0.3 MHz. To improve the achieved values, one might therefore implement a feedback mechanism that shifts the frequency of the impinging photons according to the error signal of the lock. Alternatively, improving the passive stability of the cavity, i.e. using a vibration isolation stage, might be implemented in a future setup.

References

1. E.T. Jaynes, F. W. Cummings, Comparison of quantum and semiclassical radiation theories with application to the beam maser. Proc. IEEE **51**(1), 89–109. ISSN: 0018-9219 (1963). doi: 10.1109/PROC.1963.1664
2. M. Fox, *Quantum Optics : An Introduction*. (Oxford University Press, 2006), p. 00288. ISBN: 978-0-19-152425-7
3. C.W. Gardiner, M.J. Collett, Input and output in damped quantum systems: Quantum stochastic differential equations and the master equation. Phys. Rev. A **31**(6), 3761–3774 (1985). doi: 10.1103/PhysRevA.31.3761. http://link.aps.org/doi/10.1103/PhysRevA.31.3761
4. J.I. Cirac et al., Quantum state transfer and entanglement distribution among distant nodes in a quantum network. Phys. Rev. Lett. **78**(16), 3221–3224 (1997). doi:10.1103/PhysRevLett.78.3221. http://link.aps.org/doi/10.1103/PhysRevLett.78.3221
5. D.F. Walls, G. Gerard, J. Milburn, *Quantum Optics*. (Springer, 2008). ISBN: 3-540-28574-1

6. L.-M. Duan, H.J. Kimble, Scalable photonic quantum computation through cavity-assisted interactions, Phys. Rev. Lett. **92**(12), 127902 (2004), 00427. doi:10.1103/PhysRevLett.92.127902. http://link.aps.org/doi/10.1103/PhysRevLett.92.127902
7. C.Y. Hu et al., Giant optical faraday rotation induced by a single-electron spin in a quantum dot: applications to entangling remote spins via a single photon. Phys. Rev. B **78**(8), 085307 (2008). doi:10.1103/PhysRevB.78.085307. http://link.aps.org/doi/10.1103/PhysRevB.78.085307
8. Hyochul Kim et al., A quantum logic gate between a solid-state quantum bit and a photon. Nature Photonics **7**(5), 373–377 (2013), 00010. ISSN: 1749-4885. doi:10.1038/nphoton.2013.48. http://www.nature.com/nphoton/journal/v7/n5/full/nphoton.2013.48.html
9. C. Nölleke, Quantum state transfer between remote single atoms. 00000. PhD thesis. Technische Universität München, 2013. http://mediatum.ub.tum.de/node?id=1145613
10. C. Sames et al., Antiresonance phase shift in strongly coupled cavity QED. Phys. Rev. Lett. **112**(4), 043601 (2014). doi:10.1103/PhysRevLett.112.043601. http://link.aps.org/doi/10.1103/PhysRevLett.112.043601
11. H. Specht, Einzelatom-quantenspeicher für polarisations-qubits. PhD thesis. Technische Universität München, 2010. http://mediatum.ub.tum.de/node?id=1002627
12. E.D. Black, An introduction to pound-drever-hall laser frequency stabilization. Am. J. Phys. **69**(1), 79 (2001). ISSN: 00029505. doi:10.1119/1.1286663. http://link.aip.org/link/AJPIAS/v69/i1/p79/s1&Agg=doi
13. M. Koch, Classical and quantum dynamics of a strongly coupled atom-cavity system. PhD thesis. Technische Universität München, 2011. http://mediatum.ub.tum.de/node?id=1086969

Chapter 5
Nondestructive Detection of an Optical Photon

This chapter describes the nondestructive detection of optical photons using the controlled phase gate mechanism explained in Sect. 1.2 and Chap. 4. The contents of the following sections have been published in [1]: *Nondestructive Detection of an Optical Photon*. A. Reiserer, S. Ritter, G. Rempe. Science **342**, 1349 (2013).

5.1 Introduction

The detection of optical photons is a process that is at the heart of quantum optics. So far, however, all single-photon detectors were based on absorption, thus fully destroying the photonic quantum state. In contrast, nondestructive detection, in other words the ability to watch individual photons fly by, has for decades been the "ultimate goal" [2] of optical measurements: First, it allows to detect one and the same photon repeatedly, which allows for improving the detection efficiency simply by concatenating several devices. Second, it can herald the presence of a photon without affecting unmeasured properties of its quantum state, e.g. its temporal shape or its polarization. Both implications are of high relevance for the rapidly evolving research fields of quantum measurement [3], quantum communication [4, 5], optical quantum computing [6, 7], and quantum networks [8–11].

In order to nondestructively detect optical photons, the interaction mechanism [12] described in Chap. 4 has been employed in this thesis. To this end, an incoming photon, contained in a faint laser pulse, is reflected from the cavity, in which a single atom is prepared in a superposition of two ground states. Upon reflection, the phase of the atomic superposition state is flipped. Measuring the atomic state thus allows to unambiguously detect the photon.

A detailed theoretical treatment of the atom-photon interaction mechanism is given in Chap. 4 and in [12, 13]. For an intuitive explanation, see Sect. 1.2. In the

A. Reiserer, *A Controlled Phase Gate Between a Single Atom and an Optical Photon*, Springer Theses,
DOI 10.1007/978-3-319-26548-3_5

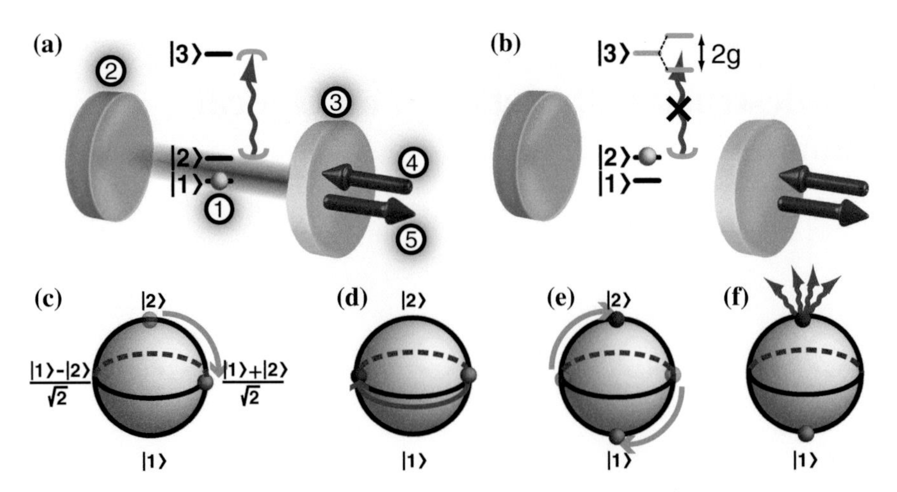

Fig. 5.1 **a, b** Sketch of the setup and atomic level scheme. A single atom, (1), is trapped in an optical cavity that consists of a high-reflector, (2), and a coupling mirror, (3). A resonant photon is impinging on, (4), and reflected off, (5), the cavity. **a** If the atom is in state $|1\rangle_a$, the photon (*red curly arrow*) enters the cavity (*blue semicircles*) before being reflected. In this process, the combined atom-photon state acquires a phase shift of π. **b** If the atom is in $|2\rangle_a$, the strong coupling on the $|2\rangle_a \leftrightarrow |3\rangle_a$ transition leads to a normal-mode splitting of $2g$, such that the photon cannot enter the cavity and is directly reflected without a phase shift. **c–f** Procedure to measure whether a photon has been reflected. **c** The atomic state, visualized on the Bloch sphere, is prepared in the superposition state $\frac{1}{\sqrt{2}}(|1\rangle_a + |2\rangle_a)$. **d** If a photon impinges, the atomic state is flipped to $\frac{1}{\sqrt{2}}(|1\rangle_a - |2\rangle_a)$. **e** The atomic state is rotated by $\frac{\pi}{2}$. **f** Fluorescence detection is used to discriminate between the states $|1\rangle_a$ and $|2\rangle_a$.(reprinted from [1])

following, the atomic level structure (Fig. 3.1) is approximated by a three-level system (see Fig. 5.1a, b). The cavity is resonant with the transition between the atomic states $|2\rangle_a$ and $|3\rangle_a$, and the state $|1\rangle_a$ is far detuned. To use the conditional phase shift for nondestructive photon detection, the atom is prepared in the superposition state $\frac{1}{\sqrt{2}}(|1\rangle_a + |2\rangle_a)$ (Fig. 5.1c), as explained in detail in Chap. 3. If there is no impinging photon, the atomic state remains unchanged (Fig. 5.1d, green filled circle). If, however, a photon is reflected, the atomic state becomes (omitting a global phase) $\frac{1}{\sqrt{2}}(|1\rangle_a + |2\rangle_a)\,|1\rangle_p \rightarrow \frac{1}{\sqrt{2}}(|1\rangle_a - |2\rangle_a)\,|1\rangle_p$ (red arrow and red filled circle). To measure this phase flip, a $\pi/2$ rotation maps the atomic state $\frac{1}{\sqrt{2}}(|1\rangle_a + |2\rangle_a)$ onto $|1\rangle_a$, while $\frac{1}{\sqrt{2}}(|1\rangle_a - |2\rangle_a)$ is rotated to $|2\rangle_a$ (Fig. 5.1e). Subsequently, cavity-enhanced fluorescence state detection [14] is used to discriminate between the atomic states $|1\rangle_a$ and $|2\rangle_a$ (Fig. 5.1f, see also Sect. 3.1). Note that two photons in the input pulse lead to a phase shift of $e^{i2\pi} = 1$. The used sequence therefore measures the odd-even parity of the photon number. As long as the average photon number per measurement interval is much smaller than one, only zero or one photon events are present and the detection result is unambiguous.

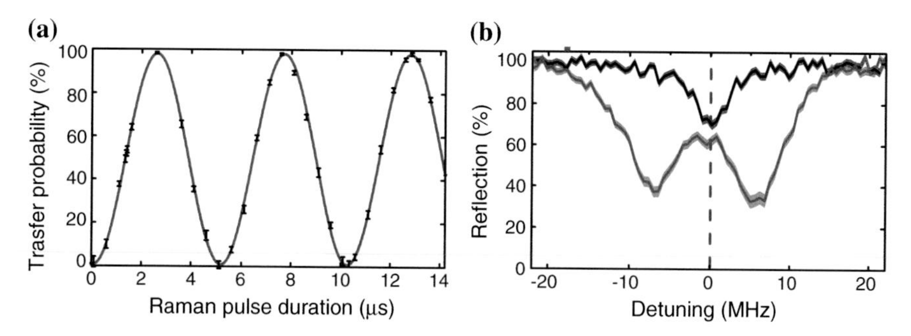

Fig. 5.2 **a** Rabi oscillations of the atomic population when the atom is prepared in $|2\rangle_a$ and two Raman laser beams are applied for a variable duration. The *red fit curve* gives a visibility of 97 %. **b** Reflection off the atom-cavity system as a function of probe laser frequency with the atom in the strongly coupled state $|2\rangle_a$ (*red*) or in the uncoupled state $|1\rangle_a$ (*black*). The statistical standard error is given by the thickness of the lines. (from [1])

5.2 Characterization Measurements

The techniques to prepare, control and read out the atomic state are described in Chap. 3. To characterize this process, Rabi oscillations are measured (Fig. 5.2a), similar to the measurements presented in Sect. 3.2.4. The observed visibility of 97 % represents an upper bound for the quality of the state preparation, rotation and readout process.

To characterize the coupling between the atom and impinging light, the reflection of the system is measured with the atom prepared in $|2\rangle_a$ as a function of the probe light frequency (red data in Fig. 5.2b), similar to the measurements in Sect. 4.1. The recorded data is in good agreement with the results presented in Fig. 4.2a; the small deviations can be explained by a slightly modified alignment of the setup. The observed normal-mode splitting demonstrates that the atom is strongly coupled. On resonance, 62(2) % of the impinging photons are reflected. With increasing coupling strength, this value is expected to approach unity. When the atom is prepared in the uncoupled state $|1\rangle_a$, 70(2) % of the incoming light is reflected on resonance (black data in Fig. 5.2b). The missing 30 % are either transmitted through the high-reflector or lost via scattering or absorption.

5.3 Nondestructive Photon Detection

Having characterized the individual steps of the protocol, they are now combined to detect photons in a nondestructive way. The atom is prepared in the superposition state $\frac{1}{\sqrt{2}}(|1\rangle_a + |2\rangle_a)$. Within a 2.5 μs long trigger interval, a weak coherent laser pulse is sent in, which has an average photon number of $\bar{n} = 0.115(11)$. Its reflection is monitored with conventional single-photon counting modules (SPCMs). Figure 5.3a

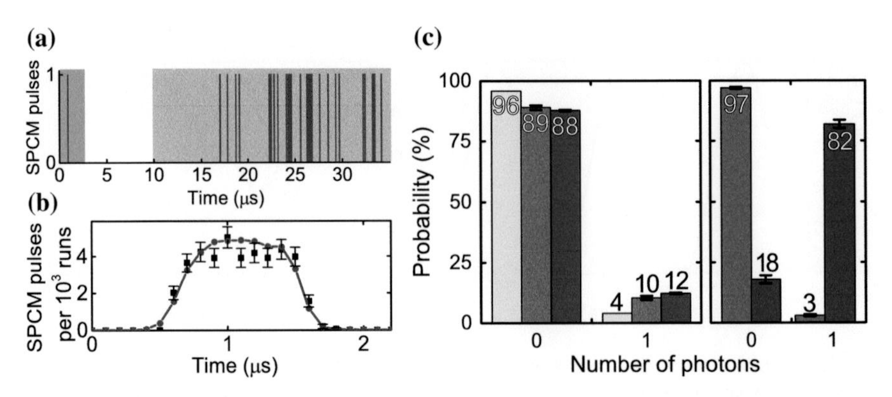

Fig. 5.3 **a** Typical trace of an experimental run. A photon (*red bar*) impinging in the trigger interval (*blue area*) leads to the emission of many photons in the readout interval (*gray area*). When the input pulse is blocked, no photons are detected in both intervals. **b** Temporal envelope of the reflected photon pulse when an atom is present (*black squares*) compared to a reference run without atom (*red points*). Within the errors, no deviation in the pulse shape is observable, except for a small amplitude change stemming from the slightly different reflectivities, see Fig. 5.2b. **c** Nondestructive detection of a single photon. The probability of detecting zero or one photon is plotted. *Yellow* Result of the SPCM detection. *Gray* Calculated input pulse, taking into account the SPCM detection efficiency. *Red* Result of the atomic state readout. *Green* Atomic state readout without impinging photon pulse. *Blue* Atomic state, conditioned on the SPCM detection of a reflected photon in the trigger interval. (from [1])

shows a typical experimental run, where a photon was subsequently detected (red line in the blue trigger interval). Therefore, after $\pi/2$ rotation of the atomic state, many fluorescence photons are observed [14] in the 25 μs long readout interval (gray), unambiguously signaling the atomic state change induced by the detected photon. Thus, in the case shown in Fig. 5.3a, a photon is detected twice: by the nondestructive detector and with a conventional, absorptive SPCM. The depicted trace also indicates that the setup works as an all-optical switch [15] which does not destroy the impinging trigger photon and also does not affect its temporal envelope. The latter can be seen in Fig. 5.3b, where the arrival-time histogram of the photons detected with the SPCMs after reflection from the setup is shown. The data taken during the nondestructive photon measurement (black squares) do not show a significant deviation from the reference curve recorded without atom (red points)—except for a small reduction in amplitude which is consistent with the results of Fig. 5.2b.

In the following, the photon detection efficiency of the nondestructive detection is investigated. The probability of detecting a photon in the input pulse is given in Fig. 5.3c. The results obtained with calibrated conventional SPCMs, without and with correction for their limited quantum efficiency of 55(5) %, are shown as yellow and gray bars, respectively. The red bars are obtained from the atomic-state readout. The probability to nondestructively detect one photon (12 %) is slightly higher than the probability to actually have one photon in the input pulse (10 %).

This is explained by a 2.9(4) % probability to observe a 'dark count': When the input pulse is blocked (green bars in Fig. 5.3c), no photons are observed, neither in the blue nor in the gray interval of Fig. 5.3a, in 97.1(4) % of all runs. In the remaining cases, many fluorescence photons are observed during the atomic state readout, corresponding to a dark count of the nondestructive photon detector. This is caused by imperfections in the atomic state preparation, rotation and readout and might be improved by magnetic shielding of the setup and by using more complex state-rotation techniques such as composite pulses [16].

Including the effect of dark counts, a comparison of the gray and red bars in Fig. 5.3c shows excellent agreement, but does not reveal information about potential systematic errors. Therefore, we also analyze correlations between the detection of a reflected photon by the SPCMs and by our nondestructive detector. The blue bars show the probability $\eta_{\text{cond}}^{\bar{n}} = 82.1(1.7)\,\%$ of finding the atom in $|2\rangle_a$, conditioned on the detection of a photon by the SPCMs. The obtained value allows to gain additional insight into the photon detection process and to derive a precise value for the unconditional detection efficiency η, as will be explained in the following Sect. 5.4. Thereafter, the mechanisms that limit the experimentally achieved value are discussed in Sect. 5.5.

5.4 Calculation of the Unconditional Detection Efficiency

To derive the unconditional detection efficiency η, it has to be considered that in the above measurements, weak coherent laser pulses with an average photon number of $\bar{n} = 0.115(11)$ are used, such that $p_1 = 10.3\,\%$ of the pulses contain a single photon, $p_2 = 0.6\,\%$ contain two and only a negligible fraction of 0.02 % contains more than two. Input states with two photons are ideally not detected by our setup, because the resulting phase shift of 2π returns the atom to its original state. Therefore, characterization of the device with coherent pulses yields a conditional detection efficiency which is systematically lower than $\eta_{\text{cond}}^{\text{n=1}}$, the value expected for single-photon input pulses. Because of the low probability for more than two photons in the input pulse, they have a negligible influence and are therefore neglected in the following. The detection of exactly one photon by the SPCM can thus be the result of four different scenarios:

The most likely, ideal situation is that the impinging pulse contains one photon, which is detected by the SPCM with a probability of $p_{\text{det}} = r\epsilon$. Here, $r = 0.66(2)$ is the probability that the photon is reflected from the cavity and $\epsilon = 0.55(5)$ is the quantum efficiency of the SPCM. The total probability for this case is $p_1 p_{\text{det}}$. In the second case, there are two photons in the impinging pulse, both of which are reflected from the cavity, but only one is detected by the SPCM. This happens with a probability of $p_2 p_{\text{det,refl}}$, with $p_{\text{det,refl}} = 2r^2\epsilon(1-\epsilon)$. In this case, the atomic phase shift is 2π and our nondestructive detection mechanism yields the incorrect result "no photon". In the third case, one of the two photons is detected, but the other is absorbed or scattered by our setup. This happens with a probability of $p_2 p_{\text{det,abs}}$, with $p_{\text{det,abs}} = 2r\epsilon(1-r)$, and projects the atom to either state $|1\rangle_a$ or state $|2\rangle_a$. The final $\pi/2$ rotation then results in an equal superposition of $|1\rangle_a$ and $|2\rangle_a$, such that our nondestructive detection mechanism leads to the incorrect result "no photon" with 50 % probability. Finally, a small fraction of 0.4 % of the SPCM detection events are caused by stray or dark counts without any impinging trigger photon. The corresponding probability for a dark count is $p_{\text{dark}} = 1.6 \cdot 10^{-4}$. Since there is no phase shift on the atom, our detector will give the result "no photon".

Considering these cases, the conditional detection efficiency of our device for single-photon input pulses is:

$$\eta_{\text{cond}}^{\text{n=1}} = \frac{p_{\text{tot}}\eta_{\text{cond}} - \frac{1}{2}p_2 p_{\text{det,abs}}}{p_1 p_{\text{det}}} = 87\,\%. \tag{5.1}$$

Here, $p_{\text{tot}} = p_{\text{dark}} + p_1 p_{\text{det}} + p_2(p_{\text{det,refl}} + p_{\text{det,abs}})$ is the total probability for a single SPCM detection event. To derive the *unconditional* single-photon detection efficiency η, two cases have to be considered: In the first case, a single impinging photon is reflected and the detection mechanism works as intended. In the second case, a single photon is impinging, but not reflected and the atomic state is thereby projected: If the photon is transmitted through the high-reflector or absorbed or scattered by the mirrors, it first had to enter the cavity, which means that the atom is in $|1\rangle_a$ with a high probability. If the photon is scattered by the atom, the latter ends up in $|2\rangle_a$. In both cases, atomic state rotation and subsequent readout give the correct result only with a probability of 50 %. Therefore, we calculate the unconditional single-photon detection efficiency of our device to be

$$\eta = r\eta_{\text{cond}}^{\text{n=1}} + (1-r)\frac{1}{2} = 74\,\%. \tag{5.2}$$

In contrast to all absorbing detectors, the efficiency of our detector can be further improved by attempting more measurements. For m concatenated devices, the efficiency to detect a single photon increases to

$$\eta \sum_{i=0}^{m-1} \left(r\,(1-\eta)\right)^i . \tag{5.3}$$

This yields 87 % for two of our nondestructive detectors, and 89 % for three or more devices. The maximum value achievable with our system is limited by absorption and scattering losses of both the atom and the cavity mirrors. To further improve, a decrease in cavity loss or an increase in atom-cavity coupling strength would be required. Both can be achieved either in Fabry-Perot [17] or other [18–20] resonators.

5.5 Analysis of Experimental Imperfections

In addition to the influence of coherent input pulses and dark counts discussed earlier, several experimental imperfections contribute to the reduction of the conditional detection efficiency η_{cond} from unity. The most severe reduction stems from the imperfect transversal spatial mode-matching of the input photons and the cavity mode (estimated to be $q = 92(2)\,\%$). As a consequence, some photons do not interact with the cavity at all. They are perfectly reflected and reach the SPCM with unit efficiency, but do not leave a trace in the atom. The photons interacting with the cavity, on the other hand, only reach the detector with probability r. Therefore, $\frac{1-q}{1-q+rq} = 12(3)\,\%$ of the photons seen by the SPCM did not interact with the cavity. The other imperfections that contribute to a reduced η_{cond} are: First, the atomic state preparation, rotation and readout are not perfect, which leads to a $3\,\%$ reduction of η_{cond}. Second, the reflection probabilities for the atom in state $|1\rangle_a$ ($70\,\%$) and $|2\rangle_a$ ($62\,\%$) are different. Therefore, reflection of the photon leaves the atom in a superposition state which is not exactly on the equator of the Bloch sphere. The estimated reduction of η_{cond} due to this effect is $0.4\,\%$. Finally, small additional reductions of η_{cond} might be caused by fluctuations of the atomic light shift and the relative stability of the laser and cavity frequencies (standard deviation$\leq$ 300 kHz, see Sect. 4.2), as well as cavity birefringence that can lead to non-circular polarization components which do not couple to the atomic state $|3\rangle_a$.

None of the imperfections has a fundamental limit. Therefore, it should be possible to further increase the efficiency achieved in our first proof-of-principle experiment, which already compares well with state-of-the-art absorbing single-photon detectors [21–23].

References

1. A. Reiserer, S. Ritter, G. Rempe, Nondestructive detection of an optical photon. Science **342**(6164), 1349–1351 (2013). doi:10.1126/science.1246164. http://www.sciencemag.org/content/342/6164/1349
2. P. Grangier, J.A. Levenson, J.-P. Poizat. Quantum non-demolition measurements in optics. Nature **396**(6711), 537–542 (1998). ISSN: 0028-0836. doi:10.1038/25059. http://www.nature.com/nature/journal/v396/n6711/abs/396537a0.html
3. H.M. Wiseman, G.J. Milburn, *Quantum Measurement and Control*, 00000. (Cambridge University Press, 2010). ISBN: 978-0-521-80442-4
4. N. Gisin et al., Quantum cryptography. Rev. Mod. Phys **74**(1), 145–195 (2002). doi:10.1103/RevModPhys.74.145. http://link.aps.org/doi/10.1103/RevModPhys.74.145
5. N. Gisin, R. Thew, Quantum communication. Nature Photonics **1**(3), 165–171 (2007), 00356. ISSN: 1749-4885. doi:10.1038/nphoton.2007.22. http://www.nature.com/nphoton/journal/v1/n3/abs/nphoton.2007.22.html
6. J.L. O'Brien. Optical Quantum Computing. Science **318**(5856), 1567–1570 (2007), 00304. doi:10.1126/science.1142892. http://www.sciencemag.org/content/318/5856/1567.abstract
7. J.L. O'Brien, A. Furusawa, J. Vuckovic, Photonic quantum technologies. Nature Photonics **3**(12), 687–695 (2009). ISSN: 1749-4885. doi:10.1038/nphoton.2009.229. http://dx.doi.org/10.1038/nphoton.2009.229

8. J.I. Cirac et al., Quantum state transfer and entanglement distribution among distant nodes in a quantum network. Phys. Rev. Lett. **78**(16), 3221–3224 (1997). doi:10.1103/PhysRevLett.78.3221. http://link.aps.org/doi/10.1103/PhysRevLett.78.3221
9. H. J. Kimble, The quantum internet. Nature **453**(7198), 1023–1030 (2008). ISSN:0028-0836. doi:10.1038/nature07127. http://dx.doi.org/10.1038/nature07127
10. L.-M. Duan, C. Monroe, Colloquium: quantum networks with trapped ions. Rev. Mod. Phys. **82**(2), 1209–1224 (2010). doi:10.1103/RevModPhys.82.1209. http://link.aps.org/doi/10.1103/RevModPhys.82.1209
11. S. Ritter et al. An elementary quantum network of single atoms in optical cavities. Nature **484**(7393) 195–200 (2012). ISSN: 0028-0836. doi:10.1038/nature11023. http://www.nature.com/nature/journal/v484/n7393/abs/nature11023.html
12. L.-M. Duan, H.J. Kimble, Scalable photonic quantum computation through cavity-assisted interactions. Phys. Rev. Lett. **92**(12), 127902 (2004), 00427. doi:10.1103/PhysRevLett.92.127902. http://link.aps.org/doi/10.1103/PhysRevLett.92.127902
13. J. Cho, H.-W. Lee, Generation of atomic cluster states through the cavity input-output process. Phys. Rev. Lett. **95**(16), 160501 (2005). doi:10.1103/PhysRevLett.95.160501. http://link.aps.org/doi/10.1103/PhysRevLett.95.160501
14. J. Bochmann et al., Lossless state detection of single neutral atoms. Phys. Rev. Lett. **104**(20), 203601 (2010), 00051. doi:10.1103/PhysRevLett.104.203601. http://link.aps.org/doi/10.1103/PhysRevLett.104.203601
15. W.C. et al., All-optical switch and transistor gated by one stored photon. Science **341**(6147), 768–770. (2013), 00029. ISSN: 0036-8075, 1095-9203. doi:10.1126/science.1238169. http://www.sciencemag.org/content/341/6147/768
16. L.M.K. Vandersypen, I.L. Chuang, NMR techniques for quantum control and computation. Rev. Mod. Phys. **76**(4), 1037–1069 (2005). doi:10.1103/RevModPhys.76.1037. http://link.aps.org/doi/10.1103/RevModPhys.76.1037
17. Y.Colombe et al., Strong atom-field coupling for Bose-Einstein condensates in an optical cavity on a chip". In: Nature **450**(7167), 272–276 (2007). ISSN: 0028-0836. doi:10.1038/nature06331. http://www.nature.com/nature/journal/v450/n7167/abs/nature06331.html
18. B. Dayan et al., A photon turnstile dynamically regulated by one atom. Science **319**(5866), 1062-1065 (2008), 00282. doi:10.1126/science.1152261. http://www.sciencemag.org/content/319/5866/1062.abstract
19. C. Junge et al., Strong coupling between single atoms and nontransversal photons. Phys. Rev. Lett. **110**(21), 213604 (2013). doi:10.1103/PhysRevLett.110.213604. http://link.aps.org/doi/10.1103/PhysRevLett.110.213604
20. J. D. Thompson et al., Coupling a single trapped atom to a nanoscale optical cavity. Science **340**(6137), 1202–1205 (2013). ISSN: 0036-8075, 1095-9203. doi:10.1126/science.1237125. http://www.sciencemag.org/content/340/6137/1202
21. R.H. Hadfield, Single-photon detectors for optical quantum information applications. Nature Photonics **3**(12), 696-705 (2009). ISSN: 1749-4885. doi:10.1038/nphoton.2009.230. http://www.nature.com/nphoton/journal/v3/n12/abs/nphoton.2009.230.html
22. M.D. Eisaman et al., Invited review article: single-photon sources and detectors. Rev. Sci. Instrum. **82**(7), 071101-25 (2011). ISSN: 00346748. doi:10.1063/1.3610677. http://rsi.aip.org/resource/1/rsinak/v82/i7/p071101_s1
23. F. Marsili et al., Detecting single infrared photons with 93% system efficiency. Nature Photonics **7**(3), 210–214 (2013). ISSN: 1749-4885.doi:10.1038/nphoton.2013.13. http://www.nature.com/nphoton/journal/v7/n3/abs/nphoton.2013.13.html

Chapter 6
A Quantum Gate Between a Flying Optical Photon and a Single Trapped Atom

After demonstration of the conditional phase shift mechanism in [1], it was also applied to demonstrate an atom-photon quantum gate, which is described in the following. The contents of this chapter have been published in [2]: *A quantum gate between a flying optical photon and a single trapped atom.* A. Reiserer, N. Kalb, G. Rempe, S. Ritter. Nature **508**, 237 (2014).

6.1 Introduction

The steady increase in control over individual quantum systems has backed the dream of quantum technologies that provide functionalities beyond any classical device. Two particularly promising applications have been explored during the past decade: First, photon-based quantum communication, which guarantees unbreakable encryption [3] but still has to be scaled to high rates over large distances. Second, quantum computation, which will fundamentally enhance computability [4] if it can be scaled to a large number of quantum bits. It was realized early on that a hybrid system of light and matter qubits [5] could solve the scalability problem of both fields—that of communication via quantum repeaters [6], that of computation via an optical interconnect between smaller quantum processors [7, 8]. To this end, the development of a robust two-qubit gate that allows to link distant computational nodes is "a pressing challenge" [8]. In this chapter, we demonstrate such a quantum gate between the spin state of a single trapped atom and the polarization state of an optical photon contained in a faint laser pulse. To demonstrate its versatility, the quantum gate is used to create atom-photon, atom-photon-photon, and photon-photon entangled states from separable input states.

In contrast to the original proposal [9], the presented implementation does not require interferometric stability, as the AC Stark shift of a linearly polarized dipole

A. Reiserer, *A Controlled Phase Gate Between a Single Atom and an Optical Photon*, Springer Theses,
DOI 10.1007/978-3-319-26548-3_6

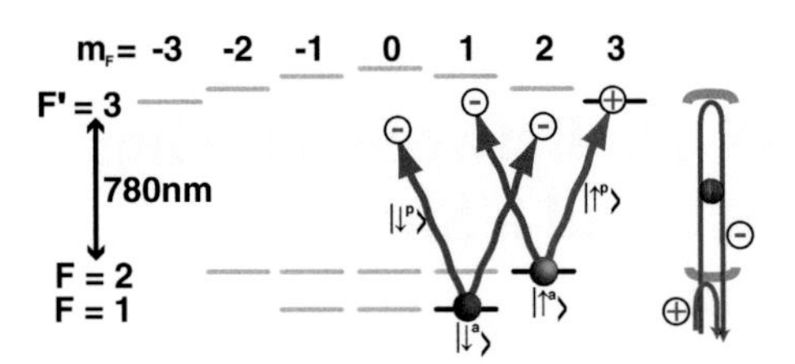

Fig. 6.1 Atomic level scheme on the D_2 line of ^{87}Rb. The photonic qubit is defined in the basis of left- ($|\downarrow^p\rangle$) and right- ($|\uparrow^p\rangle$) circular polarization. The atomic qubit is encoded in the atomic $|F, m_F\rangle$ states $|\downarrow^a\rangle \equiv |1, 1\rangle$ and $|\uparrow^a\rangle \equiv |2, 2\rangle$. Here, F denotes the atomic hyperfine state and m_F its projection onto an external magnetic field. The cavity (*blue semi-circles*) is resonant with the AC Stark-shifted $|2, 2\rangle \leftrightarrow |3, 3\rangle$ transition on the D_2 line around 780 nm. Upon reflection of a photon from the cavity, the combined atom-photon state $|\uparrow^a\uparrow^p\rangle$ (*green*, $\oplus$) acquires a phase shift of π with respect to all other states (*red*, $\ominus$). (from [2])

trap is used to split the Zeeman states of the excited atomic state manifold, as schematically depicted in Fig. 6.1. In this context, the blue-detuned trap light (see Sect. 2.2) has a negligible influence, because the atom is trapped at a node of the standing-wave light field. The red-detuned light, however, considerably shifts the frequency of the atomic transitions, depending on the polarization of the trap laser. In the experiment, π-polarized light is employed, i.e. the electric field vector is oriented parallel to the quantization axis, which coincides with the cavity axis and the direction of an externally applied magnetic field of about 0.5 G. In this configuration, the AC Stark shift is identical for all Zeeman states in the ground-state manifolds with $F = 1$ and $F = 2$. The excited state $F' = 3$, however, experiences a Zeeman-state dependent shift, as explained in Sect. 3.1.

To use these shifts for implementing a quantum gate, the impinging photon is on resonance with the transition $|2, 2\rangle \leftrightarrow |3, 3\rangle$. The transition $|2, 2\rangle \leftrightarrow |3, 1\rangle$ is thus detuned by 0.1 GHz, while all transitions from the $F = 1$ state are detuned by about 7 GHz. Therefore, only the atom in state $|\uparrow^a\rangle$ and the photon in $|\uparrow^p\rangle$ are strongly coupled (green arrow and sphere in Fig. 6.1). For all other qubit combinations (red arrows and sphere), the coupling is negligible because any atomic transition is detuned. Therefore, the reflection of a photon results in a conditional phase shift of π, i.e. a sign change, between the atomic and the photonic qubit:

$$\begin{aligned}
|\uparrow^a\uparrow^p\rangle &\rightarrow |\uparrow^a\uparrow^p\rangle \\
|\uparrow^a\downarrow^p\rangle &\rightarrow -|\uparrow^a\downarrow^p\rangle \\
|\downarrow^a\uparrow^p\rangle &\rightarrow -|\downarrow^a\uparrow^p\rangle \\
|\downarrow^a\downarrow^p\rangle &\rightarrow -|\downarrow^a\downarrow^p\rangle
\end{aligned}$$

This conditional phase shift allows to construct a universal quantum gate that can be transformed into any two-qubit gate using rotations of the individual qubits, which are implemented with wave plates for the photon and Raman transitions for the atom. With respect to the photonic basis states $|\uparrow^p_x\rangle \equiv \frac{1}{\sqrt{2}}(|\uparrow^p\rangle + |\downarrow^p\rangle)$ and $|\downarrow^p_x\rangle \equiv \frac{1}{\sqrt{2}}(|\uparrow^p\rangle - |\downarrow^p\rangle)$, the conditional phase shift represents an atom-photon controlled-NOT (CNOT) gate.

6.2 CNOT Truth Table

The action of the quantum CNOT gate is a flip of the photonic target qubit, controlled by the quantum state of the atom, similar to its classical analogue. A first step to characterize the gate is therefore to measure a classical truth table. To this end, the atomic state is prepared by optical pumping either into the uncoupled $F = 1$ states, corresponding to $|\downarrow^a\rangle$, or into the coupled $|\uparrow^a\rangle$ state (see Sect. 3.1). Subsequently, faint laser pulses (average photon number $\bar{n} = 0.3$) in $|\downarrow^p_x\rangle$ or $|\uparrow^p_x\rangle$ are reflected from the cavity and measured with single-photon counting modules in a polarization-resolving setup. To ensure spectral mode matching [9], we use a Gaussian wavepacket with a full width at half maximum (FWHM) of 0.7 μs, corresponding to a FWHM bandwidth of 0.6 MHz, which is almost an order of magnitude smaller than the cavity FWHM linewidth of 5 MHz. After the reflection process, the atomic state is measured within 3 μs using cavity-enhanced hyperfine-state detection (as described in Sect. 3.1). The results are shown in Fig. 6.2a (see also the table in Fig. 6.4a), where the bars represent the normalized probabilities to detect a certain output state for each of the orthogonal input states.

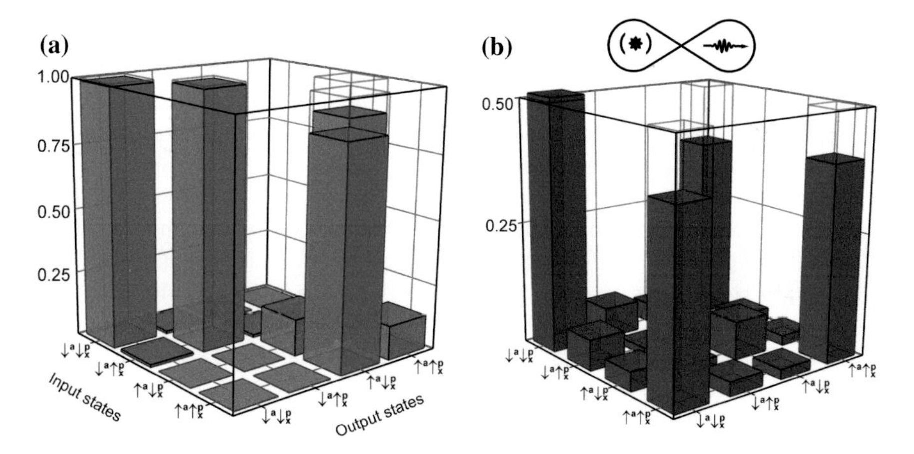

Fig. 6.2 **a** Measured truth table. The *bars* represent the normalized probability of obtaining a certain output state for a complete orthogonal set of input states. Open *blue bars* indicate the action of an ideal controlled-NOT gate. **b** Entangled atom-photon state generated via the gate operation. The *bars* show the absolute value of the density-matrix elements. The fidelity with the maximally entangled $|\Phi^+\rangle$ Bell state (open *blue bars*) is 80.7(0.5) %. The *icon* at the *top* of the figure symbolizes atom-photon entanglement. (from [2])

The control and target qubits are expected to be unchanged when the control qubit is in the state $|{\downarrow^a}\rangle$, which is accomplished with a probability of 99 %. This number is limited by imperfections in the detection of the photon polarization and the atomic hyperfine state. When the control qubit is in $|{\uparrow^a}\rangle$, the expected flip of the photonic target qubit is observed with a probability of 86 %. The statistical errors in the depicted data are negligible. However, we observe ambient-temperature-related drifts of about 2 % on a timescale of several hours. The flip probability is predominantly limited by two effects: First, by optical mode matching, because the transverse overlap between the free-space mode of the photon and the cavity mode is 92(3) %. Second, by the quality of preparing the state $|{\uparrow^a}\rangle$, which is successful with 96(1) % probability. Finally, by the relative stability between the cavity resonance and the frequency of the impinging laser pulse, which is about 300 kHz. None of these imperfections has a fundamental limit.

6.3 Atom-Photon Entanglement

The decisive feature that discriminates a quantum gate from a classical one is the generation of entangled states from separable input states. To characterize this property, faint laser pulses ($\bar{n} = 0.07$, FWHM 0.7 μs) are reflected from the setup and the evaluation is post-selected on those cases where a single photon has subsequently been detected. The input state is $|{\downarrow^a_x \downarrow^p_x}\rangle$, such that the gate generates the maximally entangled $|\Phi^+_{ap}\rangle$ state:

$$|{\downarrow^a_x \downarrow^p_x}\rangle \rightarrow |\Phi^+_{ap}\rangle = \frac{1}{\sqrt{2}}(|{\uparrow^a \uparrow^p_x}\rangle + |{\downarrow^a \downarrow^p_x}\rangle)$$

Both, the atomic and photonic qubit are measured in three orthogonal bases. This allows to reconstruct the density matrix of the combined atom-photon state using quantum-state tomography and a maximum-likelihood estimation [10]. The result is shown in Fig. 6.2b (see also the table in Fig. 6.4b). In accordance with the truth table measurement above, the density matrix is slightly asymmetric. While the value of $|{\downarrow^a \downarrow^p_x}\rangle\langle{\downarrow^a \downarrow^p_x}|$ (left corner) is close to the ideal 0.5, the elements in the other corners are smaller. The fidelity with the expected $|\Phi^+_{ap}\rangle$ state is $F_{\Phi^+_{ap}} = \langle\Phi^+_{ap}|\,\rho_{ap}\,|\Phi^+_{ap}\rangle = 80.7(0.5)\%$, where the standard error has been determined with the Monte-Carlo technique [10]. A detailed analysis of the major experimental imperfections that reduce the fidelity will be given in Sect. 6.4. In the depicted measurement, the fidelity with a slightly rotated, maximally entangled state of the form $\frac{1}{\sqrt{2}}(|{\uparrow^a \uparrow^p_x}\rangle + e^{-i\varphi}|{\downarrow^a \downarrow^p_x}\rangle)$ can be higher, probably due to a small frequency offset between the cavity and the photon. We find a maximum value of 83.0 % for $\varphi = 0.11\pi$.

In principle, the gate mechanism presented in this work is deterministic. In our experimental implementation, the photon is not back-reflected [1] from the coupled system $|{\uparrow^a \uparrow^p}\rangle$ with a probability of 34(2) % (due to the finite cooperativity

$C = \frac{g^2}{2\kappa\gamma} = 3$) and in the uncoupled cases with a probability of 30(2) % (due to the non-zero transmission of the high reflector and the mirror scattering and absorption losses, see Sect. 4.1). The small difference in reflectivity also contributes slightly (<1 %) to the observed reduction in fidelity [11]. The achieved loss level nevertheless allows for scalable quantum computation [12] and deterministic quantum state transfer [13]. One would still observe nonclassical correlations without post-selection in case a perfect single-photon source and a perfect detector were used to characterize our device. Besides, we expect that it is possible to dramatically improve the achieved value in next-generation cavities with increased atom-cavity coupling strength [14–17] and reduced losses.

6.4 Experimental Imperfections

In this section, the major experimental imperfections that reduce the fidelity of the atom-photon entangled state (Fig. 6.2) with $|\Phi^+\rangle$ are explained in more detail. As the effects are expected to be uncorrelated, they are considered independently.

The major contribution stems from the spatial mismatch between the Gaussian cavity mode and the transversal mode profile of the impinging photons. The mismatch is independently measured to be $\xi = 8(3)$ %. The unmatched fraction of the pulse will be reflected from the cavity without a phase shift, leaving both the atom and the photon in their initial state. This product state has a fidelity of $F_u = 0.25$ with the desired Bell state. The matched fraction, however, is only reflected in $\varepsilon = 69$ % of the cases. This leads to an estimated reduction of the fidelity of $\frac{\xi}{\xi+(1-\xi)\cdot\varepsilon} \cdot (1 - F_u) = 8(3)$ %.

The quality of our atomic state preparation, rotation and readout also limits the achievable fidelity. In a Ramsey spectroscopy measurement (as described in 3.2.4), we detect the atom in the expected state with a probability of 95(1) %. We therefore expect a fidelity reduction of 5(1) %.

Besides, imperfections in the photonic polarization measurement arise from two effects: From detector dark counts (3.3 % of all detection events) and from the polarizing beam splitters used in the experiment (extinction ratio about 100 : 1 in reflection, 1000 : 1 in transmission). The combined fidelity reduction of both effects is $(1 - F_u) \cdot 3.3\,\% + 0.5\,\% \simeq 3\,\%$.

Finally, the gate is characterized with faint laser pulses that have a mean photon number of $\bar{n} = 0.07$. The evaluation is conditioned on the detection of a single photon. Owing to the Poissonian photon number statistics, 4 % of these single-photon detection events are caused by higher photon-number contributions in the input. This value is calculated [1] by considering the measured absorption of the setup and quantum efficiency ($\eta = 0.6$) of the single photon counters. In about half of the two-photon cases, both photons were reflected (generating a GHZ state as explained in the main text) but only one was detected. Then, the detected atom-photon state would be classically correlated with $F_c = 0.5$, giving a total reduction of the entanglement fidelity of $F_c \cdot 2\,\% = 1\,\%$. In the other half of the cases, one of the two photons was absorbed. Thus, the atomic quantum state is partially projected: While the uncoupled $|\downarrow^p\rangle$ polarization component would leave the atom in the initial state, the coupled $|\uparrow^p\rangle$ component would fully project the atomic state. Therefore, the resulting fidelity reduction for an absorbed photon is $(1 - F_u) \cdot 0.5 \cdot 2\,\% \simeq 1\,\%$. The overall fidelity reduction due to two-photon components is thus 2 %.

6.5 Atom-Photon-Photon and Photon-Photon Entanglement

The demonstrated quantum gate also allows one to generate entangled cluster states that consist of a trapped atom and several flying photons, complementary to experiments with flying atoms and trapped microwave photons [18]. To this end, the gate is applied to the photons contained in two sequentially impinging laser pulses (temporal distance 3 μs). Post-selecting events where one photon was detected in each of the input pulses, a maximally entangled Greenberger-Horne-Zeilinger (GHZ) state [19] is expected:

$$|\downarrow_x^a \downarrow_x^p \downarrow_x^p\rangle \rightarrow |\mathrm{GHZ}\rangle = \frac{1}{\sqrt{2}}(|\uparrow^a \uparrow_x^p \uparrow_x^p\rangle - |\downarrow^a \downarrow_x^p \downarrow_x^p\rangle)$$

The density matrix of the generated quantum state, again reconstructed using quantum state tomography and a maximum-likelihood estimation, is shown in Fig. 6.3a (see also the table in Fig. 6.4c). The fidelity with the ideal state $|\mathrm{GHZ}\rangle$ (open blue bars in Fig. 6.3a) is 61(2) %, proving genuine three-particle (atom-photon-photon) entanglement. The reasons for a non-unity fidelity are analogous to the case of two particles. Again, we experimentally find a higher fidelity of 67 % with the slightly rotated GHZ state $\frac{1}{\sqrt{2}}(|\uparrow^a \uparrow_x^p \uparrow_x^p\rangle - e^{-i\varphi}\,|\downarrow^a \downarrow_x^p \downarrow_x^p\rangle)$, with $\varphi = 0.21\pi$.

Finally, we investigate whether the presented gate mechanism can mediate a photon-photon interaction for optical quantum computing [9]. We employ a quantum eraser protocol [20, 21] which should allow to create a maximally entangled state out of two separable input photons. To this end, the state $|\mathrm{GHZ}\rangle$ is generated as described above and a $\frac{\pi}{2}$ rotation is applied to the atom, which transforms the state to:

$$\frac{1}{\sqrt{2}}\left[|\uparrow^a\rangle\,(|\uparrow_x^p \uparrow_x^p\rangle - |\downarrow_x^p \downarrow_x^p\rangle) - |\downarrow^a\rangle\,(|\uparrow_x^p \uparrow_x^p\rangle + |\downarrow_x^p \downarrow_x^p\rangle)\right]$$

Subsequent measurement of the atomic state disentangles the atom, which results in a maximally entangled two-photon state: If the atom is found in $|\downarrow^a\rangle$ ($|\uparrow^a\rangle$), the resulting state is $|\Phi_{pp}^+\rangle$ ($|\Phi_{pp}^-\rangle$), respectively. In the experiment, the two-photon density matrices are again reconstructed with the maximum-likelihood technique (Fig. 6.3b, see also the tables in Fig. 6.4d, e). This gives a fidelity with the expected Bell states of 67(2) % (64(2) %) for the $|\Phi_{pp}^+\rangle$ ($|\Phi_{pp}^-\rangle$) state. The achieved values proof photon-photon entanglement. Their small difference can be explained by the fact that a detection of the atom in $F = 1$ selects only those events where it has

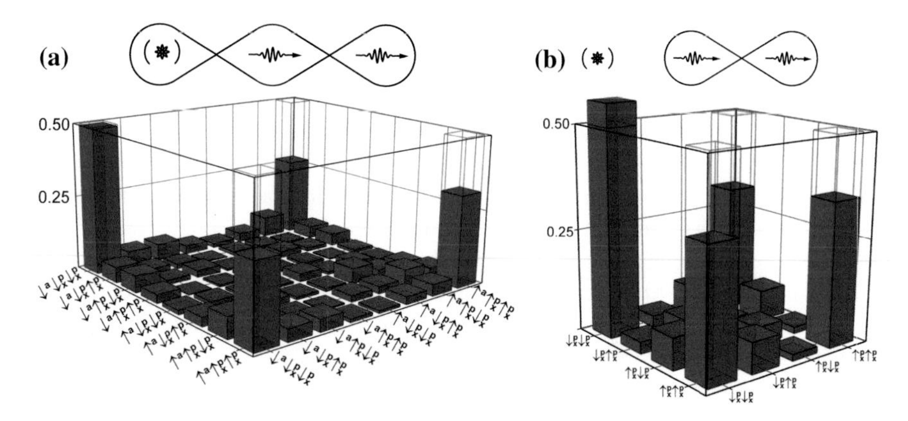

Fig. 6.3 **a** Entangled state between one atom and two photons, generated by reflecting two faint laser pulses from the cavity. The *bars* show the absolute value of the reconstructed density matrix elements. The fidelity with the maximally entangled state $|\mathrm{GHZ}\rangle$ is 61(2) %. The matrix elements of $|\mathrm{GHZ}\rangle$ are depicted as open *blue bars*. The *icon* at the top of the figure symbolizes entanglement between an atom and two photons. **b** Entangled photon-photon state generated via consecutive interaction with the atom. The *bars* show the absolute value of the density-matrix elements. The fidelity with the maximally entangled $|\Phi^+\rangle$ Bell state (*open blue bars*) is 67(2) %. The *icon* at the top of the figure symbolizes photon-photon entanglement. (from [2])

a

	$\downarrow^a\downarrow^p_x$	$\downarrow^a\uparrow^p_x$	$\uparrow^a\downarrow^p_x$	$\uparrow^a\uparrow^p_x$
$\downarrow^a\downarrow^p_x$	0.989	0.008	0.003	0.000
$\downarrow^a\uparrow^p_x$	0.010	0.987	0.000	0.003
$\uparrow^a\downarrow^p_x$	0.002	0.005	0.139	0.855
$\uparrow^a\uparrow^p_x$	0.003	0.001	0.870	0.126

b

	$\downarrow^a\downarrow^p_x$	$\downarrow^a\uparrow^p_x$	$\uparrow^a\downarrow^p_x$	$\uparrow^a\uparrow^p_x$
$\downarrow^a\downarrow^p_x$	0.510	h.c.	h.c.	h.c.
$\downarrow^a\uparrow^p_x$	0.059 - 0.020 i	0.016	h.c.	h.c.
$\uparrow^a\downarrow^p_x$	- 0.039 - 0.002 i	- 0.005 - 0.019 i	0.077	h.c.
$\uparrow^a\uparrow^p_x$	0.353 - 0.131 i	0.026 - 0.010 i	- 0.011 + 0.017 i	0.398

c

	$\downarrow^a\downarrow^p_x\downarrow^p_x$	$\downarrow^a\downarrow^p_x\uparrow^p_x$	$\downarrow^a\uparrow^p_x\downarrow^p_x$	$\downarrow^a\uparrow^p_x\uparrow^p_x$	$\uparrow^a\downarrow^p_x\downarrow^p_x$	$\uparrow^a\downarrow^p_x\uparrow^p_x$	$\uparrow^a\uparrow^p_x\downarrow^p_x$	$\uparrow^a\uparrow^p_x\uparrow^p_x$
$\downarrow^a\downarrow^p_x\downarrow^p_x$	0.494	h.c.	h.c.	h.c.	h.c.	h.c.	h.c.	h.c.
$\downarrow^a\downarrow^p_x\uparrow^p_x$	0.053 - 0.004 i	0.021	h.c.	h.c.	h.c.	h.c.	h.c.	h.c.
$\downarrow^a\uparrow^p_x\downarrow^p_x$	0.054 - 0.015 i	0.013 + 0.008 i	0.025	h.c.	h.c.	h.c.	h.c.	h.c.
$\downarrow^a\uparrow^p_x\uparrow^p_x$	- 0.007 - 0.027 i	- 0.001 + 0.001 i	0.003 - 0.006 i	0.009	h.c.	h.c.	h.c.	h.c.
$\uparrow^a\downarrow^p_x\downarrow^p_x$	0.006 + 0.012 i	0.013 - 0.003 i	0.007 - 0.007 i	- 0.007 + 0.001 i	0.028	h.c.	h.c.	h.c.
$\uparrow^a\downarrow^p_x\uparrow^p_x$	0.020 - 0.022 i	0.010 - 0.005 i	0.021 + 0.001 i	- 0.007 + 0.006 i	0.010 + 0.002 i	0.054	h.c.	h.c.
$\uparrow^a\uparrow^p_x\downarrow^p_x$	0.060 - 0.035 i	0.004 + 0.001 i	+ 0.012 + 0.002 i	- 0.002 + 0.004 i	- 0.020 + 0.008 i	0.023 - 0.009 i	0.057	h.c.
$\uparrow^a\uparrow^p_x\uparrow^p_x$	- 0.212 + 0.163 i	- 0.028 + 0.030 i	- 0.006 + 0.040 i	- 0.004 + 0.010 i	- 0.005 - 0.008 i	0.017 + 0.019 i	- 0.029 + 0.025 i	0.311

d

	$\downarrow^p_x\downarrow^p_x$	$\downarrow^p_x\uparrow^p_x$	$\uparrow^p_x\downarrow^p_x$	$\uparrow^p_x\uparrow^p_x$
$\downarrow^p_x\downarrow^p_x$	0.556	h.c.	h.c.	h.c.
$\downarrow^p_x\uparrow^p_x$	0.013 + 0.025 i	0.054	h.c.	h.c.
$\uparrow^p_x\downarrow^p_x$	- 0.076 - 0.003 i	0.024 + 0.008 i	0.042	h.c.
$\uparrow^p_x\uparrow^p_x$	0.218 - 0.222 i	0.074 - 0.004 i	0.016 - 0.001 i	0.348

e

	$\downarrow^p_x\downarrow^p_x$	$\downarrow^p_x\uparrow^p_x$	$\uparrow^p_x\downarrow^p_x$	$\uparrow^p_x\uparrow^p_x$
$\downarrow^p_x\downarrow^p_x$	0.538	h.c.	h.c.	h.c.
$\downarrow^p_x\uparrow^p_x$	0.112 - 0.020 i	0.087	h.c.	h.c.
$\uparrow^p_x\downarrow^p_x$	0.066 - 0.045 i	0.068 + 0.003 i	0.073	h.c.
$\uparrow^p_x\uparrow^p_x$	- 0.223 + 0.044 i	- 0.009 + 0.045 i	- 0.016 + 0.059 i	0.303

Fig. 6.4 Numerical values of the truth table and density matrices. h.c. denotes the Hermitian conjugate. (from [2]) **a** Data of the truth table measurement depicted in Fig. 6.2a. **b** Atom-photon density matrix. The absolute values of the elements are depicted in Fig. 6.2b. **c** Atom-photon-photon density matrix. The absolute value of the elements are depicted in Fig. 6.3a. **d** Photon-photon density matrix, postselected on the detection of the atomic $|\uparrow^a\rangle$ state. The absolute value of the elements are depicted in Fig. 6.3b. **e** Photon-photon density matrix, postselected on the detection of the atomic $|\downarrow^a\rangle$ state.

initially been prepared in the correct state $|\uparrow^a\rangle$, rather than in another state of the $F = 2$ hyperfine manifold. Again, we find a higher fidelity of maximally 76 % with a rotated $|\Phi^+_{pp}\rangle$ state with $\varphi = 0.25\pi$.

The above measurements demonstrate the versatility of the presented gate mechanism and its ability to mediate a photon-photon interaction. To this end, intermediate storage of the two photons during the time required to rotate and read out the atomic state (about 3 μs) is required, which can be implemented with an optical fiber of less than one kilometer length. Conditioned on the state of the atom, the polarization of the photons then has to be rotated, e.g. using an electro-optical modulator. As an alternative to the eraser-scheme employed in this work, the first photon could be reflected from the cavity a second time [9].

References

1. A. Reiserer, S. Ritter, G. Rempe, Nondestructive detection of an optical photon. Science **342**(6164), 1349–1351 (2013). doi:10.1126/science.1246164. http://www.sciencemag.org/content/342/6164/1349
2. A. Reiserer et al., A quantum gate between a flying optical photon and a single trapped atom. Nature **508**(7495), 237–240 (2014). 00010, ISSN: 0028-0836. doi:10.1038/nature13177. http://www.nature.com/nature/journal/v508/n7495/full/nature13177.html. Accessed 9 Apr 2014
3. N. Gisin et al., Quantum cryptography. Rev. Mod. Phys. **74**(1), 145–195 (2002). doi:10.1103/RevModPhys.74.145. http://link.aps.org/doi/10.1103/RevModPhys.74.145
4. T.D. Ladd et al., Quantum computers. Nature **464**(7285), 45–53 (2010). 00859, ISSN: 0028-0836. doi:10.1038/nature08812. http://dx.doi.org/10.1038/nature08812
5. L.-M. Duan, C. Monroe, Colloquium: quantum networks with trapped ions. Rev. Mod. Phys. **82**(2), 1209–1224 (2010). doi:10.1103/RevModPhys.82.1209. http://link.aps.org/doi/10.1103/RevModPhys.82.1209
6. H.-J. Briegel et al., Quantum repeaters: the role of imperfect local operations in quantum communication. Phys. Rev. Lett. **81**(26), 5932–5935 (1998). 01541,10.1103/PhysRevLett.81.5932. http://link.aps.org/doi/10.1103/PhysRevLett.81.5932
7. C. Monroe, J. Kim, Scaling the ion trap quantum processor. Science **339**(6124), 1164–1169 (2013). ISSN: 0036-8075, 1095-9203. doi:10.1126/science.1231298. http://www.sciencemag.org/content/339/6124/1164
8. D.D. Awschalom et al., Quantum spintronics: engineering and manipulating atom-like spins in semiconductors. Science **339**(6124), 1174–1179 (2013). 00072, ISSN: 0036-8075, 1095-9203. doi:10.1126/science.1231364. http://www.sciencemag.org/content/339/6124/1174
9. L.-M. Duan, H.J. Kimble, Scalable photonic quantum computation through cavity-assisted interactions. Phys. Rev. Lett. **92**(12), 127902 (2004). 00427, doi:10.1103/PhysRevLett.92.127902. http://link.aps.org/doi/10.1103/PhysRevLett.92.127902
10. M. Paris, J. Řeháček, *Quantum State Estimation*. (Springer, 2004). ISBN:978-3-540-22329-0
11. J. Cho, H.-W. Lee, Generation of atomic cluster states through the cavity input-output process. Phys. Rev. Lett. **95**(16), 160501 (2005). doi:10.1103/PhysRevLett.95.160501. http://link.aps.org/doi/10.1103/PhysRevLett.95.160501
12. L.-M. Duan, R. Raussendorf, Efficient quantum computation with probabilistic quantum gates. Phys. Rev. Lett. **95**(8), 080503 (2005). doi:10.1103/PhysRevLett.95.080503. http://link.aps.org/doi/10.1103/PhysRevLett.95.080503
13. S.J. van Enk, J.I. Cirac, P. Zoller, Photonic channels for quantum communication. Science **279**(5348), 205-208 (1998). ISSN: 0036-8075, 1095-9203. doi:10.1126/science.279.5348.205. http://www.sciencemag.org/content/279/5348/205

14. Y. Colombe et al., Strong atom-field coupling for Bose-Einstein condensates in an optical cavity on a chip. Nature **450**(7167), 272–276 (2007). ISSN: 0028-0836. doi:10.1038/nature06331. http://www.nature.com/nature/journal/v450/n7167/abs/nature06331.html
15. B. Dayan et al., A photon turnstile dynamically regulated by one atom. Science **319**(5866), 1062–1065 (2008). 00282, doi:10.1126/science.1152261. http://www.sciencemag.org/content/319/5866/1062.abstract
16. C. Junge et al., Strong coupling between single atoms and nontransversal photons. Phys. Rev. Lett. **110**(21), 213604 (2013). doi:10.1103/PhysRevLett.110.213604. http://link.aps.org/doi/10.1103/PhysRevLett.110.213604
17. J.D. Thompson et al., Coupling a single trapped atom to a nanoscale optical cavity. Science **340**(6137), 1202–1205 (2013). ISSN: 0036-8075, 1095-9203. doi:10.1126/science.1237125. http://www.sciencemag.org/content/340/6137/1202
18. A. Rauschenbeutel et al., Step-by-step engineered multiparticle entanglement. Science **288**(5473), 2024–2028 (2000). ISSN: 0036-8075, 1095-9203. doi:10.1126/science.288.5473.2024. http://www.sciencemag.org/content/288/5473/2024
19. D.M. Greenberger, M.A. Horne, A. Zeilinger, in *Going beyond Bell's Theorem*. Bell's Theorem, Quantum Theory and Conceptions of the Universe. Fundamental Theories of Physics 37. (Springer, Netherlands, 1989), pp. 69–72. ISBN: 978-90-481-4058-9. http://link.springer.com/chapter/10.1007/978-94-017-0849-4_10
20. C.F. Roos et al., Control and measurement of three-qubit entangled states. Science **304**(5676), 1478–1480 (2004). ISSN: 0036-8075, 1095-9203. doi:10.1126/science.1097522. http://www.sciencemag.org/content/304/5676/1478
21. C.Y. Hu, W.J. Munro, J.G. Rarity, Deterministic photon entangler using a charged quantum dot inside a microcavity. Phys. Rev. B **78**(12), 125318 (2008). 00083, doi:10.1103/PhysRevB.78.125318. http://link.aps.org/doi/10.1103/PhysRevB.78.125318

Chapter 7
Summary and Outlook

In this thesis, the long standing goal [1–11] of full control over the position and motion of an atom trapped in a cavity has been achieved by implementing a three-dimensional optical lattice within the cavity. This allows, first, to deterministically localize the atom at the center of the intra-cavity field by shifting the standing-wave pattern of the lattice beams [12, 13]. This is indispensable to reproducibly achieve a constant coupling to the cavity field. Second, in this configuration Raman lasers can be used for sideband cooling to the three-dimensional motional ground state, which guarantees a constant light shift for the optically trapped atom. In this way, the ideal CQED situation, which is also assumed in most proposals on the subject, has been realized [14]: A point-like atom that is trapped at a fixed position within the field of an overcoupled cavity in the strong coupling regime.

This is an ideal starting point for the cavity-based generation of nonclassical states of motion [15], the transfer of quantum states between atomic motion and light [16], and the observation of numerous optomechanical effects [17] with single phonons and single photons [18, 19]. In addition, the photon emission and absorption efficiencies and fidelities in coherent quantum networks, which we also demonstrated in the course of this thesis [20–22], will no longer be limited by the atomic motion. Moreover, the exquisite localization of the atom facilitates the implementation of proposals that require both constant coupling strength and optical phase stability, such as the generation of entangled states of several atoms in one cavity [23–27]. Finally, the presented advances in atom cooling and trapping have enabled the implementation of a novel interaction mechanism [28], which is based on photon reflection from a resonant cavity.

This mechanism has been employed in this thesis for the nondestructive detection of an optical photon [29] and for the implementation of an atom-photon quantum gate [30]. This lays the ground for numerous future experiments. A first step is the repeated nondestructive measurement of a single optical photon. Next, with a higher number of photons in the impinging laser pulse, the odd-even parity measurement presented in [29] allows one to generate new quantum states of optical light fields, e.g. Schrödinger-cat states [31]. Moreover, the presented quantum gate [30] can

A. Reiserer, *A Controlled Phase Gate Between a Single Atom and an Optical Photon*, Springer Theses,
DOI 10.1007/978-3-319-26548-3_7

be further extended to an entangling gate between several successively impinging photons [28] or between several atoms trapped in the same or even in remote cavities, thus efficiently generating atomic cluster states [32–34]. Besides, universal quantum computation in a decoherence-free subspace has been proposed [35]. Finally, the gate mechanism can be used for quantum communication using a redundant quantum parity code [36], or to perform a deterministic optical Bell-state measurement [37]. This would dramatically increase the efficiency of teleportation between remote atoms [22, 38, 39] and therefore the prospects for the implementation of a quantum repeater [40] and a quantum network [21, 41] on a global scale [42].

References

1. D. Vernooy, H. Kimble, Well-dressed states for wave-packet dynamics in cavity QED. Phys. Rev. A **56**(5), 4287–4295 (1997). ISSN: 1050-2947, 1094-1622. doi:10.1103/PhysRevA.56.4287. http://authors.library.caltech.edu/3240/
2. J. Ye, D.W. Vernooy, H.J. Kimble, Trapping of single atoms in cavity QED. Phys. Rev. Lett. **83**(24), 4987–4990 (1999). doi:10.1103/PhysRevLett.83.4987. http://link.aps.org/doi/10.1103/PhysRevLett.83.4987
3. P.W.H. Pinkse et al., Trapping an atom with single photons. Nature **404**(6776), 365–368 (2000). ISSN: 0028-0836. doi:10.1038/35006006. http://dx.doi.org/10.1038/35006006
4. T. Fischer et al., Feedback on the motion of a single atom in an optical cavity. Phys. Rev. Lett. **88**(16), 163002 (2002), 00106. doi:10.1103/PhysRevLett.88.163002. http://link.aps.org/doi/10.1103/PhysRevLett.88.163002
5. J. McKeever et al., State-insensitive cooling and trapping of single atoms in an optical cavity. Phys. Rev. Lett. **90**(13), 133602 (2003), 00324. doi:10.1103/PhysRevLett.90.133602. http://link.aps.org/doi/10.1103/PhysRevLett.90.133602
6. P. Maunz et al., Cavity cooling of a single atom. Nature **428**(6978), 50–52 (2004). ISSN: 0028-0836. doi:10.1038/nature02387. http://dx.doi.org/10.1038/nature02387
7. S. Nußmann et al., Vacuum-stimulated cooling of single atoms in three dimensions. Nature Phys. **1**(2), 122–125 (2005). ISSN: 1745-2473. doi:10.1038/nphys120. http://www.nature.com/nphys/journal/v1/n2/abs/nphys120.html
8. A.D. Boozer et al., Cooling to the ground state of axial motion for one atom strongly coupled to an optical cavity. Phys. Rev. Lett. **97**(8), 083602 (2006). doi:10.1103/PhysRevLett.97.083602. http://link.aps.org/doi/10.1103/PhysRevLett.97.083602
9. A. Kubanek et al., Photon-by-photon feedback control of a single-atom trajectory. Nature **462**(7275), 898–901 (2009). ISSN: 0028-0836. doi:10.1038/nature08563. http://dx.doi.org/10.1038/nature08563
10. M. Koch et al., Feedback cooling of a single neutral atom. Phys. Rev. Lett. **105**(17), 173003 (2010), 00034. doi:10.1103/PhysRevLett.105.173003. http://link.aps.org/doi/10.1103/PhysRevLett.105.173003
11. T. Kampschulte et al., Electromagnetically-induced-transparency control of single-atom motion in an optical cavity. Phys. Rev. A **89**(3), 033404 (2014), 00000. doi:10.1103/PhysRevA.89.033404. http://link.aps.org/doi/10.1103/PhysRevA.89.033404
12. S. Kuhr et al., Deterministic delivery of a single atom. Science **293**(5528), 278–280 (2001). ISSN: 0036-8075, 1095-9203. doi:10.1126/science.1062725. http://www.sciencemag.org/content/293/5528/278
13. S. Nußmann et al., Submicron positioning of single atoms in a microcavity. Phys. Rev. Lett. **95**(17), 173602 (2005). doi:10.1103/PhysRevLett.95.173602. http://link.aps.org/doi/10.1103/PhysRevLett.95.173602

14. A. Reiserer et al., Ground-state cooling of a single atom at the center of an optical cavity. Phys. Rev. Lett. **110**(22), 223003 (2013). doi:10.1103/PhysRevLett.110.223003. http://link.aps.org/doi/10.1103/PhysRevLett.110.223003
15. H. Zeng, F. Lin, Quantum conversion between the cavity fields and the center-of-mass motion of ions in a quantized trap. Phys. Rev. A **50**(5), R3589–R3592 (1994), 00074. doi:10.1103/PhysRevA.50.R3589. http://link.aps.org/doi/10.1103/PhysRevA.50.R3589
16. A.S. Parkins, H. J. Kimble, Quantum state transfer between motion and light. J. Opt. B: Quantum and Semiclassical Opt. **1**(4), 496–504 (1999). ISSN: 1464-4266, 1741-3575. doi:10.1088/1464-4266/1/4/323. http://iopscience.iop.org/1464-4266/1/4/323
17. M. Aspelmeyer, T.J. Kippenberg, F. Marquardt, Cavity optomechanics. Rev. Mod. Phys. **86**(4), 1391–1452 (2014), 00002. doi:10.1103/RevModPhys.86.1391. http://link.aps.org/doi/10.1103/RevModPhys.86.1391
18. P. Rabl, Photon blockade effect in optomechanical systems. Phys. Rev. Lett. **107**(6), 063601 (2011). doi:10.1103/PhysRevLett.107.063601. http://link.aps.org/doi/10.1103/PhysRevLett.107.063601
19. A. Nunnenkamp, K. Børkje, S.M. Girvin, Single-photon optomechanics. Phys. Rev. Lett. **107**(6), 063602 (2011), 00129. doi:10.1103/PhysRevLett.107.063602. http://link.aps.org/doi/10.1103/PhysRevLett.107.063602
20. H.P. Specht et al., A single-atom quantum memory. Nature **473**(7346), 190–193 (2011), 00128. ISSN: 0028-0836. doi:10.1038/nature09997. http://dx.doi.org/10.1038/nature09997
21. S. Ritter et al., An elementary quantum network of single atoms in optical cavities. Nature **484**(7393), 195–200 (2012). ISSN: 0028-0836. doi:10.1038/nature11023. http://www.nature.com/nature/journal/v484/n7393/abs/nature11023.html
22. C. Nölleke et al., Efficient teleportation between remote single-atom quantum memories. Phys. Rev. Lett. **110**(14), 140403 (2013), 00034. doi:10.1103/PhysRevLett.110.140403. http://link.aps.org/doi/10.1103/PhysRevLett.110.140403
23. T. Pellizzari et al., Decoherence, continuous observation, and quantum computing: a cavity QED model. Phys. Rev. Lett. **75**(21), 3788–3791 (1995). doi:10.1103/PhysRevLett.75.3788. http://link.aps.org/doi/10.1103/PhysRevLett.75.3788
24. Anders S. Sørensen, K. Mølmer, Measurement induced entanglement and quantum computation with atoms in optical cavities. Phys. Rev. Lett. **91**(9), 097905 (2003). doi:10.1103/PhysRevLett.91.097905. http://link.aps.org/doi/10.1103/PhysRevLett.91.097905
25. A.S. Sørensen, K. Mølmer, Probabilistic generation of entanglement in optical cavities. Phys. Rev. Lett. **90**(12), 127903 (2003), 00063. doi:10.1103/PhysRevLett.90.127903. http://link.aps.org/doi/10.1103/PhysRevLett.90.127903
26. M.J. Kastoryano, F. Reiter, A.S. Sørensen, Dissipative preparation of entanglement in optical cavities. Phys. Rev. Lett. **106**(9), 090502 (2011), 00096. doi:10.1103/PhysRevLett.106.090502. http://link.aps.org/doi/10.1103/PhysRevLett.106.090502
27. G. Nikoghosyan, M.J. Hartmann, M. B. Plenio, Generation of mesoscopic entangled states in a cavity coupled to an atomic ensemble. Phys. Rev. Lett. **108**(12), 123603 (2012). doi:10.1103/PhysRevLett.108.123603. http://link.aps.org/doi/10.1103/PhysRevLett.108.123603
28. L.-M. Duan, H.J. Kimble, Scalable photonic quantum computation through cavity-assisted interactions. Phys. Rev. Lett. **92**(12), 127902 (2004), 00427. doi:10.1103/PhysRevLett.92.127902. http://link.aps.org/doi/10.1103/PhysRevLett.92.127902
29. A. Reiserer, S. Ritter, G. Rempe, Nondestructive detection of an optical photon. Science **342**(6164), 1349–1351 (2013). doi:10.1126/science.1246164. http://www.sciencemag.org/content/342/6164/1349
30. A. Reiserer et al., A quantum gate between a flying optical photon and a single trapped atom. Nature **508**(7495), 237–240 (2014), 00010. ISSN: 0028-0836. doi:10.1038/nature13177. http://www.nature.com/nature/journal/v508/n7495/full/nature13177.html
31. B. Wang, L.-M. Duan, Engineering superpositions of coherent states in coherent optical pulses through cavity-assisted interaction. Phys. Rev. A **72**(2), 022320 (2005). doi:10.1103/PhysRevA.72.022320. http://link.aps.org/doi/10.1103/PhysRevA.72.022320

32. Y.-F. Xiao et al., Realizing quantum controlled phase flip through cavity QED. Phys. Rev. A **70**(4), 042314 (2004). doi:10.1103/PhysRevA.70.042314. http://link.aps.org/doi/10.1103/PhysRevA.70.042314
33. J. Cho, H.-W. Lee, Generation of atomic cluster states through the cavity input-output process. Phys. Rev. Lett. **95**(16), 160501 (2005). doi:10.1103/PhysRevLett.95.160501. http://link.aps.org/doi/10.1103/PhysRevLett.95.160501
34. L.-M. Duan, B. Wang, H. J. Kimble, Robust quantum gates on neutral atoms with cavity-assisted photon scattering. Phys. Rev. A **72**(3), 032333 (2005), 00155. doi:10.1103/PhysRevA.72.032333. http://link.aps.org/doi/10.1103/PhysRevA.72.032333
35. P. Xue, Y.-F. Xiao, Universal quantum computation in decoherence-free subspace with neutral atoms. Phys. Rev. Lett. **97**(14), 140501 (2006), 00061. doi:10.1103/PhysRevLett.97.140501. http://link.aps.org/doi/10.1103/PhysRevLett.97.140501
36. W.J. Munro et al., Quantum communication without the necessity of quantum memories. en. Nature Photonics **6**(11), 777–781 (2012). ISSN: 1749-4885. doi:10.1038/nphoton.2012.243. http://www.nature.com/nphoton/journal/v6/n11/full/nphoton.2012.243.html
37. C. Bonato et al., CNOT and bell-state analysis in the weak-coupling cavity QED regime. Phys. Rev. Lett. **104**(16), 160503 (2010), 00112. doi:10.1103/PhysRevLett.104.160503. http://link.aps.org/doi/10.1103/PhysRevLett.104.160503
38. S. Olmschenk et al., Quantum teleportation between distant matter qubits. Science **323**(5913), 486–489 (2009). doi:10.1126/science.1167209. http://www.sciencemag.org/content/323/5913/486.abstract
39. C. Nölleke, Quantum state transfer between remote single atoms. 00000. PhD thesis. Technische Universität München, 2013. http://mediatum.ub.tum.de/node?id=1145613
40. H.-J. Briegel et al., Quantum repeaters: the role of imperfect local operations in quantum communication. Phys. Rev. Lett. **81**(26), 5932–5935 (1998), 01541. doi:10.1103/PhysRevLett.81.5932. http://link.aps.org/doi/10.1103/PhysRevLett.81.5932
41. L.-M. Duan, C. Monroe, Colloquium: quantum networks with trapped ions. Rev. Mod. Phys. **82**(2), 1209–1224 (2010). doi:10.1103/RevModPhys.82.1209. http://link.aps.org/doi/10.1103/RevModPhys.82.1209
42. H.J. Kimble, The quantum internet. Nature **453**(7198), 1023–1030 (2008). ISSN:0028-0836. doi:10.1038/nature07127. http://dx.doi.org/10.1038/nature07127